Root Temperature and Plant Growth— A Review

This and other publications of the
Commonwealth Agricultural Bureaux
can be obtained through any major bookseller
or direct from:
Commonwealth Agricultural Bureaux,
Central Sales,
Farnham Royal,
Slough SL2 3BN,
England

ROOT TEMPERATURE AND PLANT GROWTH

A REVIEW

by

A. J. COOPER, B.Sc., Ph.D.

Glasshouse Crops Research Institute,
Littlehampton, Sussex

Research Review No. 4
Commonwealth Bureau of Horticulture and Plantation Crops,
East Malling, Maidstone, Kent

COMMONWEALTH AGRICULTURAL BUREAUX

First published in 1973 by the
Commonwealth Agricultural Bureaux,
Farnham Royal, Slough, SL2 3BN, England. £1.60

SBN 85198 271 9

Contents

INTRODUCTION

This review is divided into four parts. In the first part data are presented which show the influence of root temperature on some aspects of plant growth and development, namely dry weight gain, plant height, root extent and form, leaf expansion, leaf form, flowering and fruiting. The second part is a similar presentation of data illustrating the influence of root temperature on the processes of photosynthesis, respiration, water absorption and transpiration. This is followed in Part 3 by an examination of naturally occurring soil temperatures, which leads to a consideration of the effects of a change in root temperature relative to the influence of constant root temperatures. The final part is an examination of the practical steps that can be taken to control root temperature during crop production.

PART 1

THE INFLUENCE OF ROOT TEMPERATURE ON PLANT GROWTH

Many of the data presented in the following thirteen sections have been obtained experimentally from comparisons made between different root temperatures which have been maintained constant throughout the experiments. The presentation is largely factual with little speculation. The speculation is made here, however, that extrapolation from these data (obtained with constant root temperatures) which might suggest the response of plant growth in the fluctuating root temperatures of commercial crop production, might need to be treated with caution.

1.1 Dry weight gain by the whole plant

The response to root temperature of the dry weight of strawberry plants can be seen in Figure 1, drawn from data collected by Proebsting (181) from two

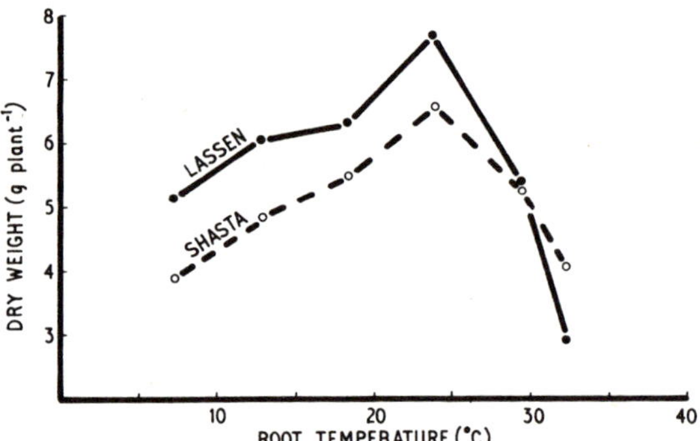

Figure 1: Influence of root temperature on the dry weights of two cultivars of strawberry plants.

cultivars grown for three months at root temperatures between 7 and 32°C at a common aerial temperature. There was an optimal temperature near 25°C, and the temperature dependence below the optimum was less marked than that above it. The data suggest that the two cultivars responded differently to root temperature.

Nelson and Tukey (155) found different responses to root temperature when they grew cultivars of apple rootstocks (M.1, M.2, M.7, M.9 and M.16) at root temperatures of 7, 13, 19 and 25°C. The dry weight gain of all the rootstocks was low at 7°C but it increased with the root temperature. M.2 and M.9 were the most responsive to root temperature, having an optimum near 13°C and ceasing to grow above a root temperature between 13 and 19°C. The rootstocks M.1 and M.7 were less responsive, having optima at 13 and 19°C, respectively, and showing a less rapid decline in dry weight above the optimal temperature. M.16 was the least responsive; its dry weight increased markedly between 7 and 13°C, and continued to increase slightly with further rise in root temperature to an optimum beyond 25°C.

Differences in root temperature response have also been found between species and between genera. Figure 2 has been drawn from the data of Smoliak and Johnston (212), who grew 12 grasses for 90 days after seedling emergence at root temperatures of 7, 13, 18 and 27°C. None of the grasses achieved its greatest dry weight at 7°C but the continuous lines in Figure 2

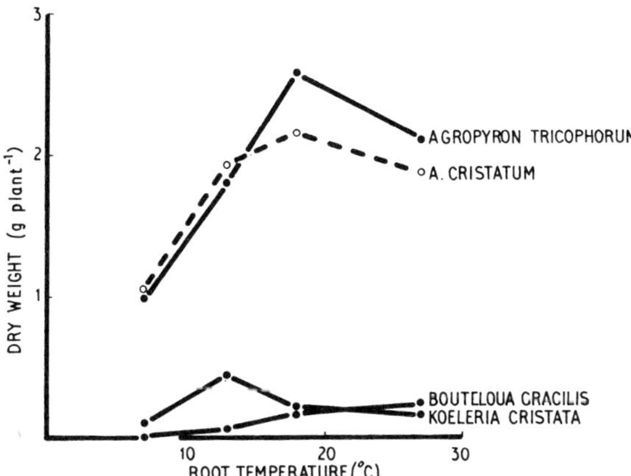

Figure 2: Influence of root temperature on the dry weights of four grasses.

illustrate the observed differences in response. *Koeleria cristata* had an optimal root temperature near 13°C, *Agropyron tricophorum* near 18°C and *Bouteloua gracilis* grew fastest at a root temperature probably above 27°C.

Differences were also observed between species within the same genus, as shown by the discontinuous and dotted lines in Figure 2 which suggest that the optimal root temperature for *Agropyron cristatum* is lower than that for *A. tricophorum.*

Seedlings of *Tilia americana* have a greater dry weight at higher root temperatures within the range 10-24°C (8), as do plants of maize, *Gossypium hirsutum, Phaseolus vulgaris,* and soybeans over the range 14-26°C (232). Thus, in general, there is an increase in the dry weight of the whole plant with rising root temperature, to an optimum temperature above which dry weight declines. The change in dry weight with unit change in root temperature above the optimum has a slope which differs from that below the optimum. These changes and the values of the optimal temperature differ between genera, species and cultivars. The different responses above and below the optimum suggest that root temperature may be affecting the dry weight gain of the whole plant through different mechanisms on either side of the optimal temperature.

1.2. Dry weight gain by the shoot

The influence of root temperature on the dry weight of the shoot has been determined for many kinds of plant. In Figures 3, 4 and 5 a selection is presented from the data of Shanks and Laurie (198) on roses, Erickson and Smith (71) on guayule, Jones and Tisdale (103) on soybeans, Johnson and Hartman (102) on tobacco, Boxall (25) and Fujishige and Sugiyama (76) on tomatoes, Darrow (60) on ***Poa pratensis,*** Brouwer (34) on peas, Stuckey (216) on colonial bent grass, Wort (244) on spring wheat, Roberts and Kenworthy (190) on strawberries, Adams (2) on white pine, Grobbelaar (85) on maize, Nagai and Matsushita (152) on rice, Boxall (27) on cucumbers and Proebsting (180) on peaches. It is apparent from these data that the optimal root temperature for shoot dry weight differs between genera.

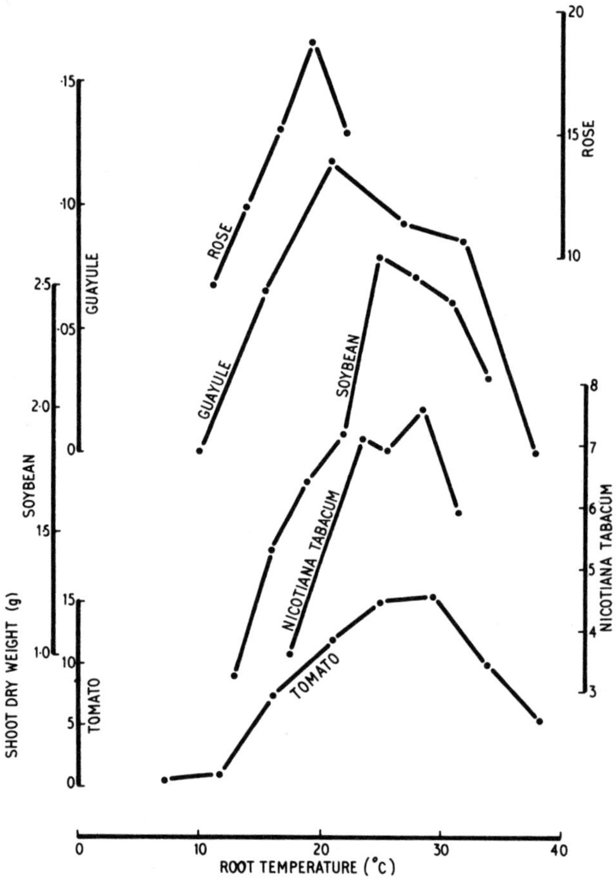

Figure 3: Influence of root temperature on shoot dry weight.

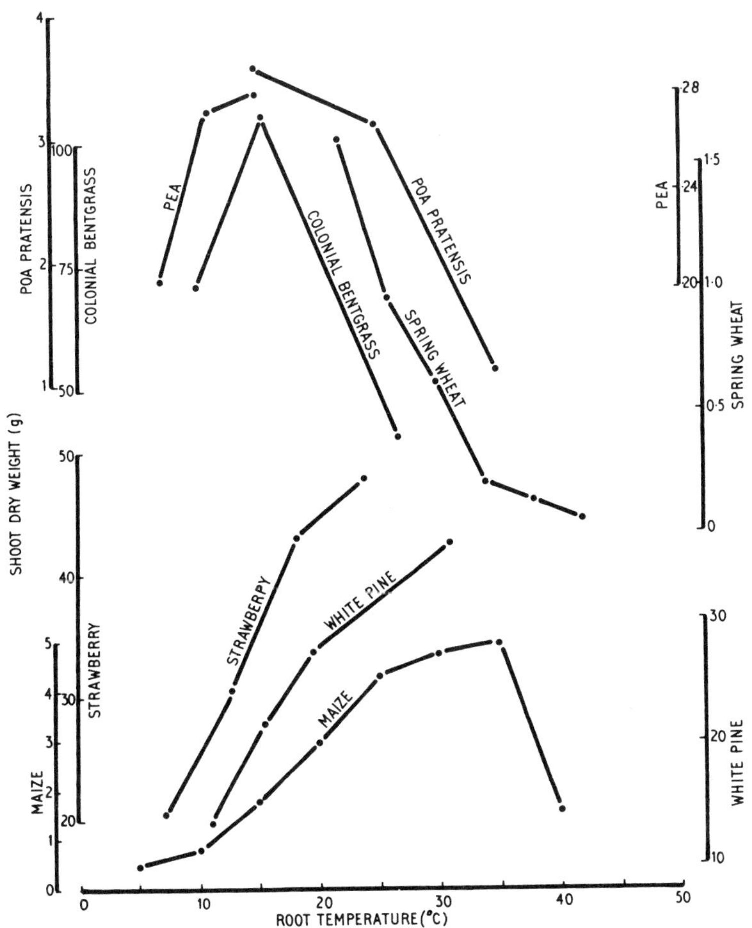

Figure 4: Influence of root temperature on shoot dry weight.

The work of Heinricks and Nielsen (90) on *Medicago* spp. showed that species within the same genus may also differ in their response to root temperature. This is illustrated in Figure 6 (left), where the average shoot dry weights of several cultivars of three *Medicago* spp. are plotted against root temperature. It was also observed that cultivars of the same species may differ in their response, as is shown in Figure 6 (right) illustrating the influence of root temperature on the dry weights of the shoots of selected cultivars of *M. falcata* and *M. media*. Dibbern (64) also reported that root temperature had different effects on the shoot dry weights of four clones of *Bromus inermis* (Figure 7).

Based on the above data a generalization about the influence of root

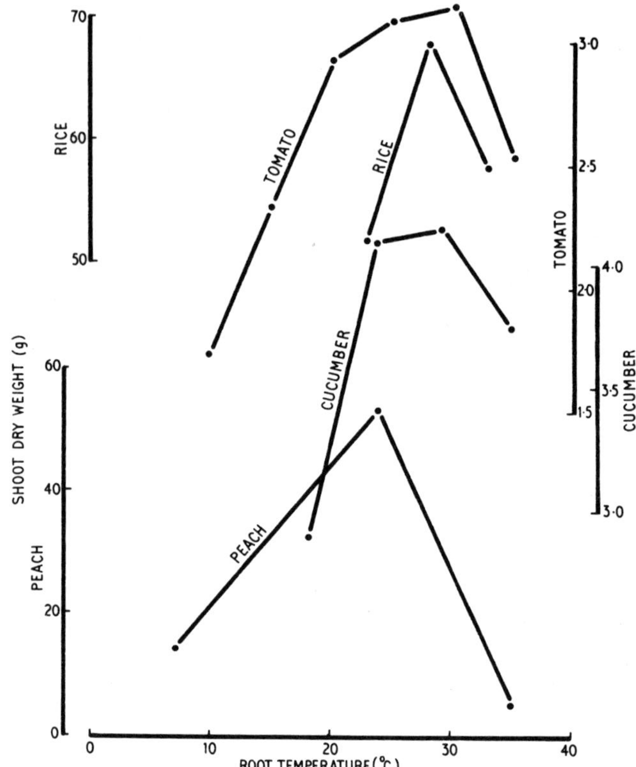

Figure 5: Influence of root temperature on shoot dry weight.

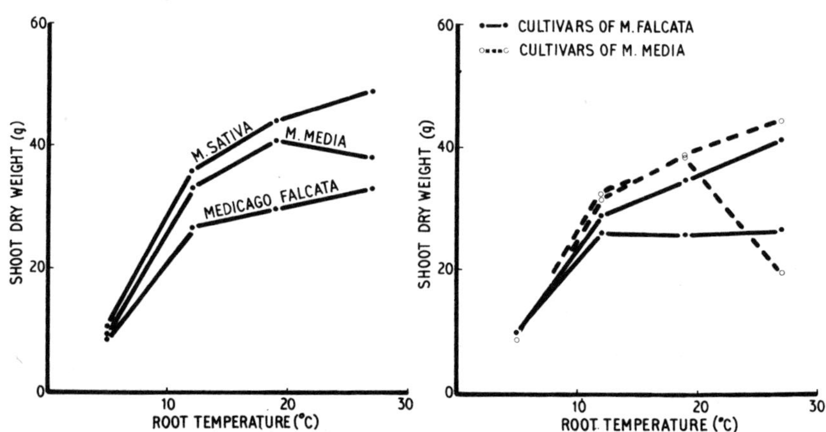

Figure 6: Influence of root temperature on shoot dry weight of three *Medicago* species (left) and of two cultivars of *M. falcata* and *M. media*.

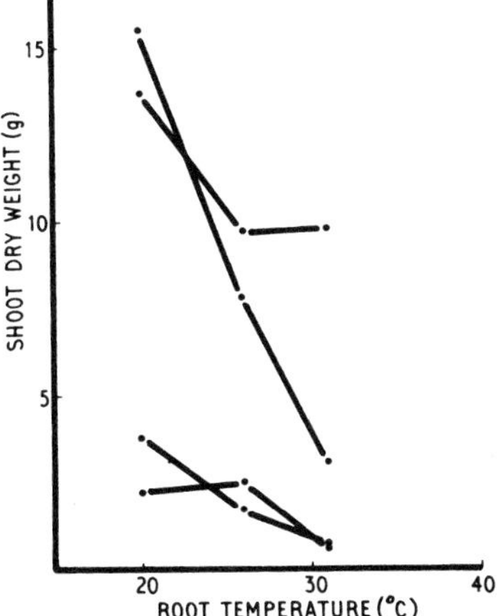

Figure 7: Influence of root temperature on shoot dry weight of four clones of *Bromus inermis*.

temperature on shoot dry weight might take the form shown in Figure 8. This suggests that:

1. The response curve below the optimal temperature for most rapid growth may be sigmoid, for example, with tomatoes and tobacco (Figure 3), maize, white pine and strawberries (Figure 4), and tomatoes and cucumbers (Figure 5).
2. The response curve above the optimum may also be sigmoid, for example guayule (Figure 3), *Poa pratensis* and spring wheat (Figure 4).

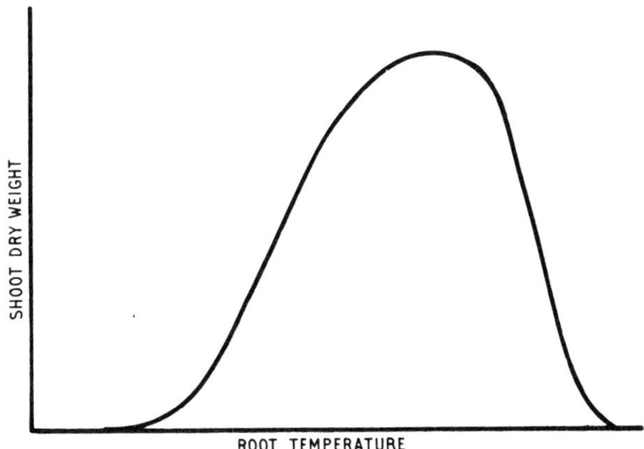

Figure 8: Diagrammatic, generalized influence of root temperature on shoot dry weight.

3. The change in shoot dry weight with unit change in root temperature above the optimum is steeper than below it, possibly because different mechanisms limit dry weight gain above and below the optimal root temperature.

4. There is an optimal band of root temperature over which temperature change has a relatively small effect on shoot dry weight; it is approximately 25-30°C for tomatoes, cucumbers, tobacco and soybeans (Figures 3 and 5), and approximately 30-35°C for maize (Figure 4).

These generalizations are supported by the very comprehensive work of Brouwer (36), who determined the dry weights of the shoots of flax, strawberries, peas, broad beans, red kidney beans, rape, maize and oats at eight root temperatures between 5 and 40°C. He expressed the dry weights of

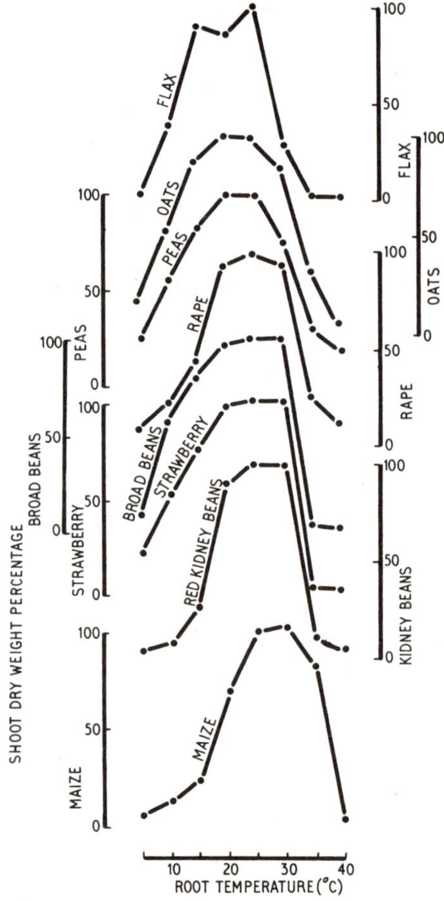

Figure 9: Shoot dry weight at several root temperatures expressed as % dry weight at 25°C.

the shoots as percentages of the dry weight at 25°C, and Figure 9 has been drawn from these data.

1.3. Dry weight gain of roots

The influence of root temperature on root dry weight is illustrated in Figures 10, 11 and 12 drawn from the work of Brouwer and Vliet (39) on peas, Erickson and Smith (71) and Benedict (18) on guayule, Meyer and Tukey (145) on *Forsythia intermedia* and *Taxus media,* Nielsen, Halstead, MacLean, Bourget and Holmes (159), Grobbelaar (85) and Knoll, Lathwell and Brady (116) on maize, Nagai and Matsushita (152) on rice, Proebsting (180) on peaches, Wort (244) on wheat, Nielsen *et al.* (159) on brome grass, Darrow (60) on *Poa pratensis,* Galligar (78) on cotton and sunflowers, Stuckey (216) on colonial bent grass, Shanks and Laurie (198, 199) on roses, Nielson and Cunningham (158) on Italian rye grass, and Nielson, Halstead,

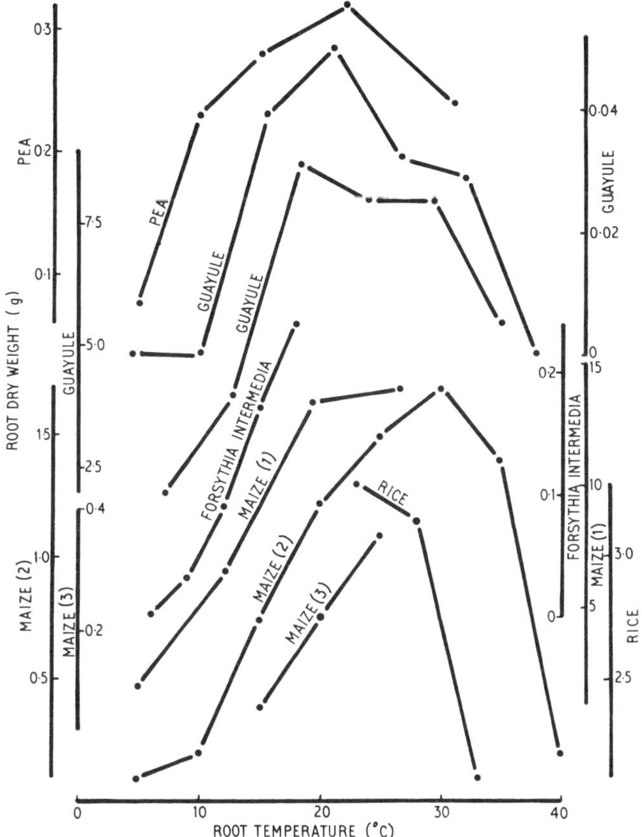

Figure 10: Effect of root temperature on root dry weight.

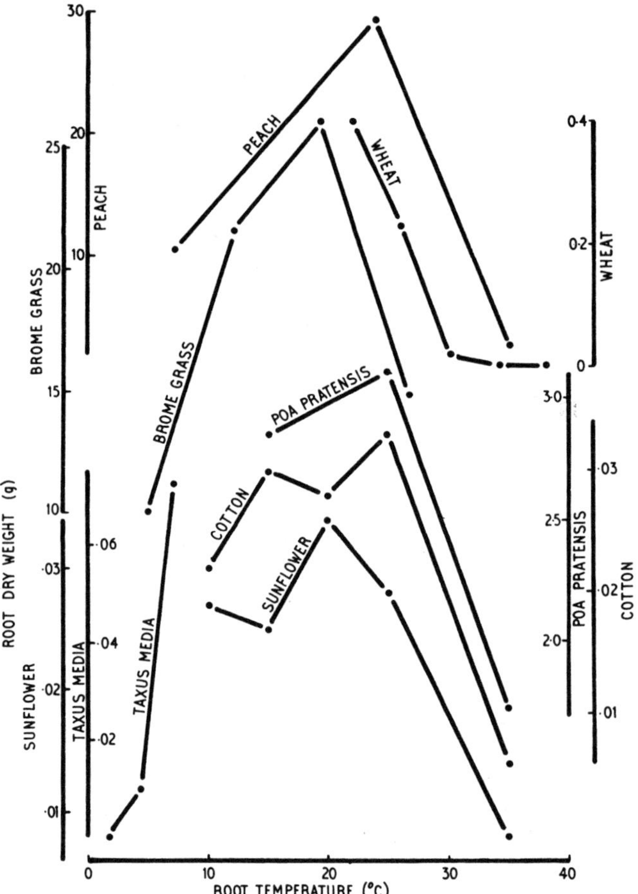

Figure 11: Effect of root temperature on root dry weight.

MacLean, Holmes and Bourget (160) and Fulton (77) on oats. It is apparent that plants in different genera respond differently to the influence of root temperature on root dry weight.

The work of Heinricks and Nielsen (90) with lucerne, illustrated in Figure 13 (left), showed that the root dry weights of plants in different species of the same genus may be differently affected by root temperature. That cultivars within the same species may also respond differently to root temperature is apparent from the data of Walker (230) on onions and of Johnson and Hartman (102) on tobacco (Figure 14). Clones within the same species may also show similar variations, as can be seen in Figure 13 (right) drawn from the data of Dibbern (64).

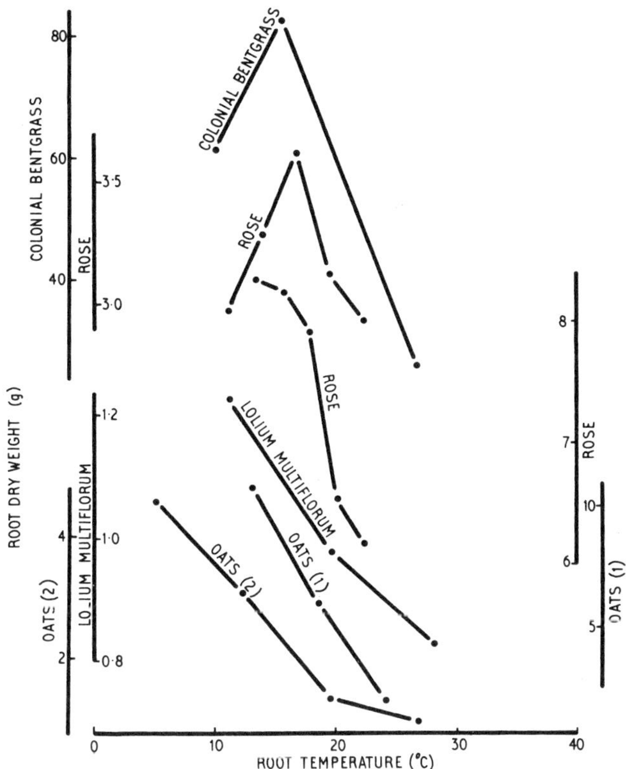

Figure 12: Effect of root temperature on root dry weight.

An examination of Figures 10, 11 and 12 suggests that the response of root dry weight to root temperature is similar in form to the generalized response curve for shoot growth of Figure 8. This conclusion is supported by the work of Brouwer (36), who determined the influence of root temperature on the roots of flax, strawberries, peas, broad beans, red kidney beans, rape, maize and oats. He expressed the dry weights as percentages of the dry weight at 25°C, and Figure 15 has been drawn from these data.

1.4. Root:shoot ratio

Although the root:shoot ratio provides information about the interrelation of the root and the shoot, it gives no information about the quantities involved, and it is quite possible that with a high root:shoot ratio the growth rate of the shoot can be greater than with a low root:shoot ratio. A low root:shoot ratio merely indicates that the partitioning of dry matter between the shoot and the root has been such that a greater proportion of the total dry matter is found in the shoot.

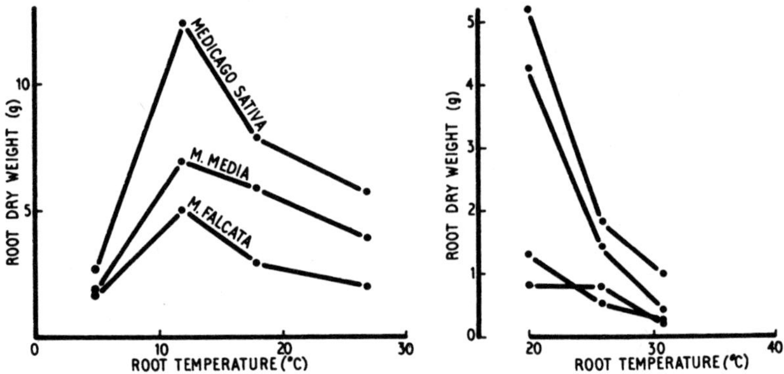

Figure 13: Effect of root temperature on root dry weight of *Medicago* species (left) and *Bromus inermis* clones (right).

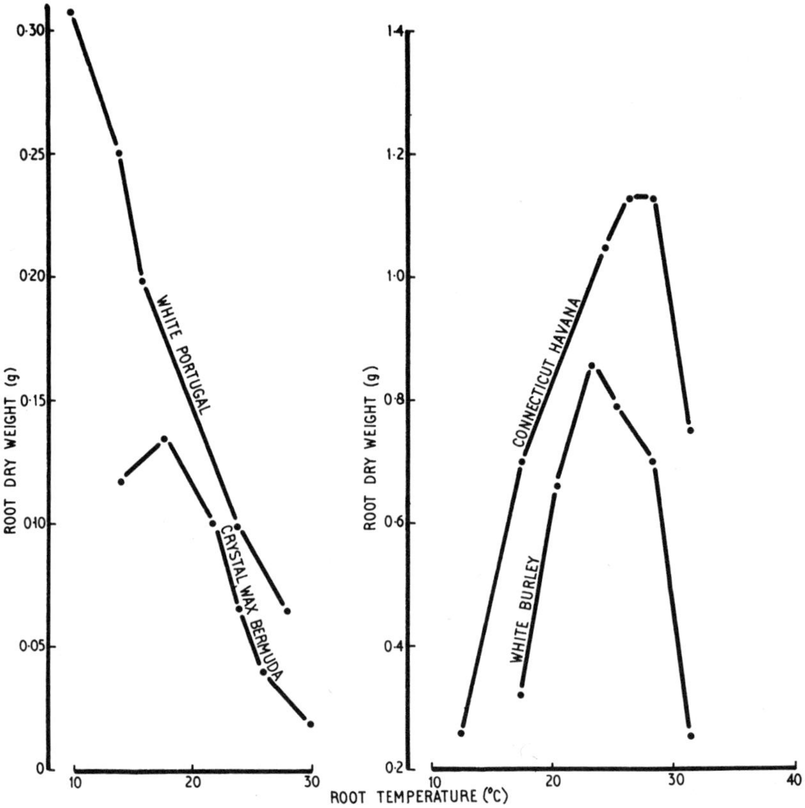

Figure 14: Effect of root temperature on root dry weights of onion cultivars (left); *Nicotiana tabacum* cultivars (right).

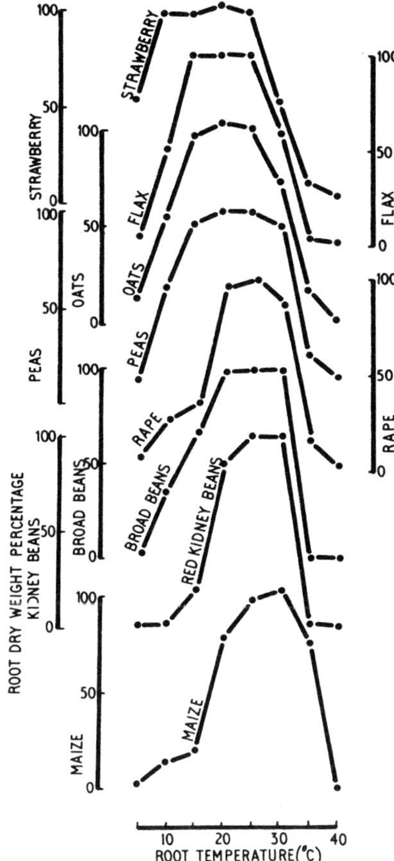

Figure 15: Root dry weights at several root temperatures expressed as % dry weight at 25°C.

That root temperature affects this partitioning between root and shoot can be seen in Figure 16 drawn from the work of Davidson (63) on *Dactylis glomerata, Bothriochloa ambigua, Festuca arundinacea* and *Lolium perenne,* Nielsen, Halstead, MacLean, Holmes and Bourget (160) on oats, Proebsting (181) on strawberries, Wort (244) on spring wheat, Nagai and Matsushita (152) on rice, Dibbern (64) on *Bromus inermis* and Grobbelaar (85) on maize. These observations suggest that there is more than one type of response of the root:shoot ratio to root temperature. The most common response is well shown by *Festuca arundinacea,* which has a high root:shoot ratio at both low and high root temperatures, whereas at intermediate root temperatures a greater proportion of the total dry matter is found in the

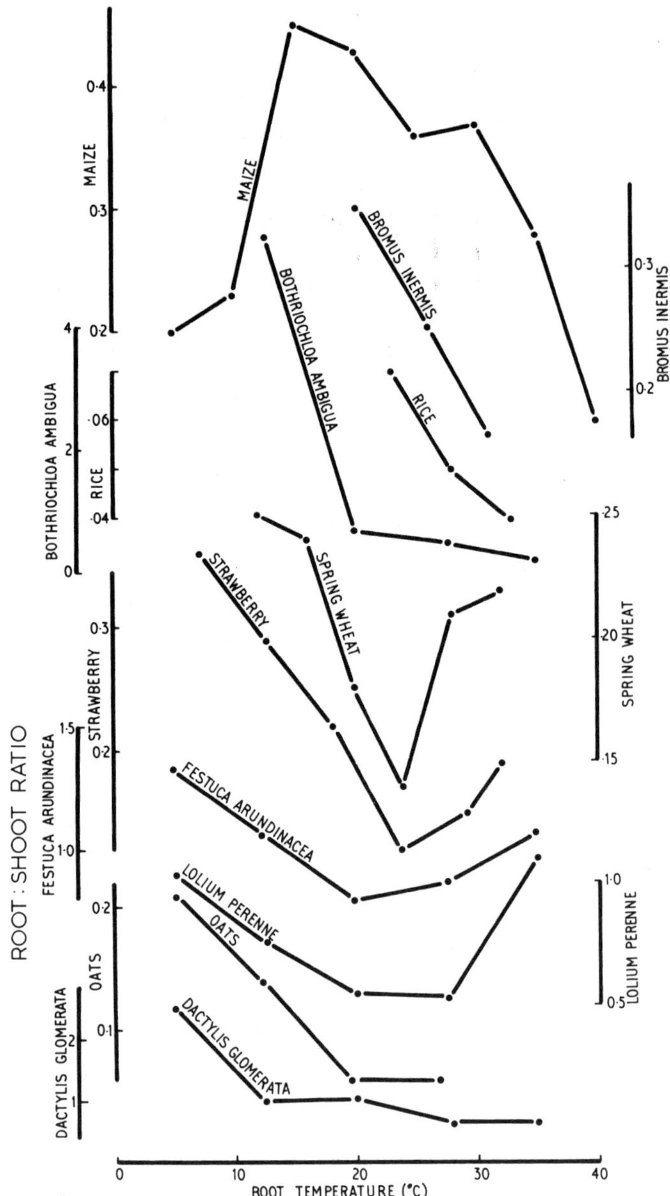

Figure 16: The effect of root temperature on the root:shoot ratio.

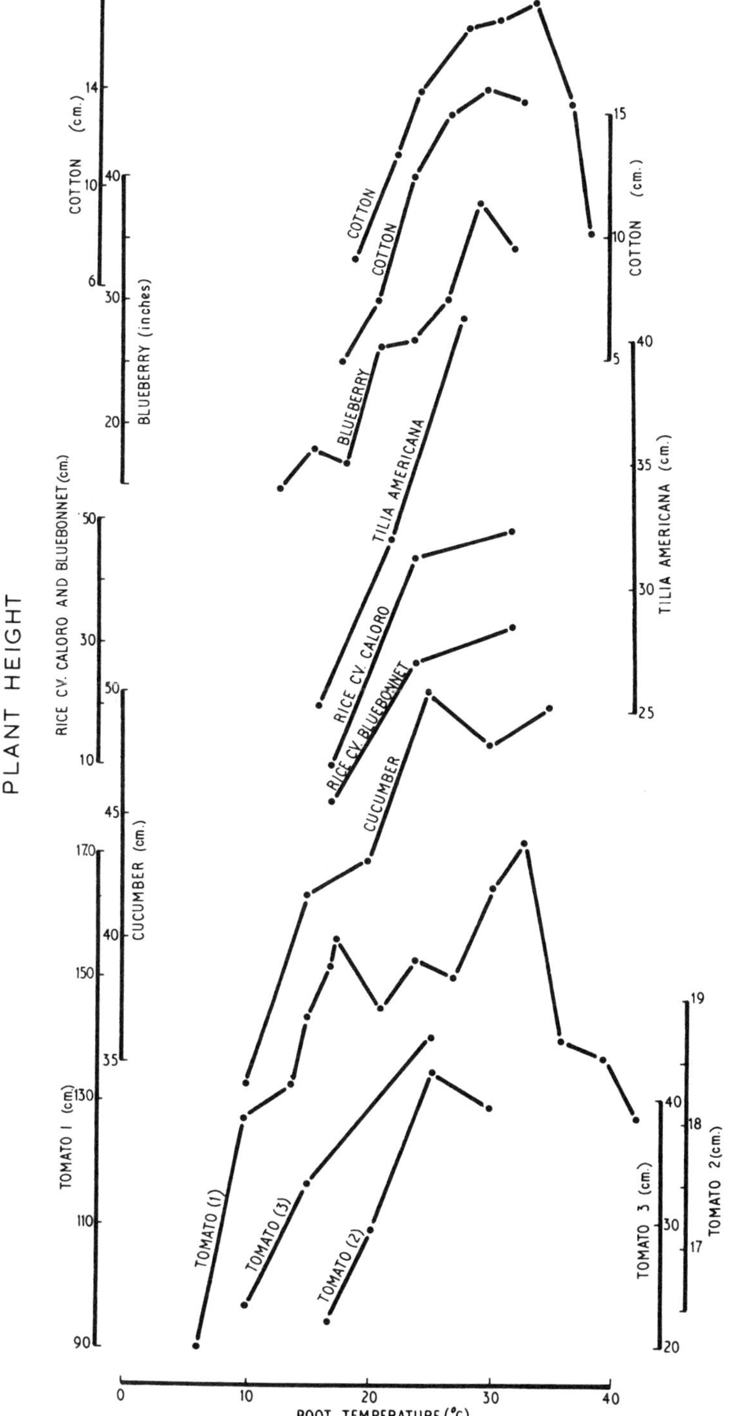

Figure 17: Influence of root temperature on plant height.

shoot. The opposite response is shown by maize, which has its highest root:shoot ratio at intermediate root temperatures, whereas at low and high root temperatures a smaller proportion of the total dry matter is found in the roots.

1.5. Plant height

The influence of root temperature on plant height has been studied in many species, but, despite this, the whole of the response curve has never been defined. One of the most complete definitions was made by Riethmann (188) for tomatoes. Figures 17 and 18, illustrating the effect of root temperature on plant height, have been drawn from these data, and those of Abd El Rahman, Kuiper and Bierhuizen (1) for tomatoes, Fujishige and Sugiyama (76) for

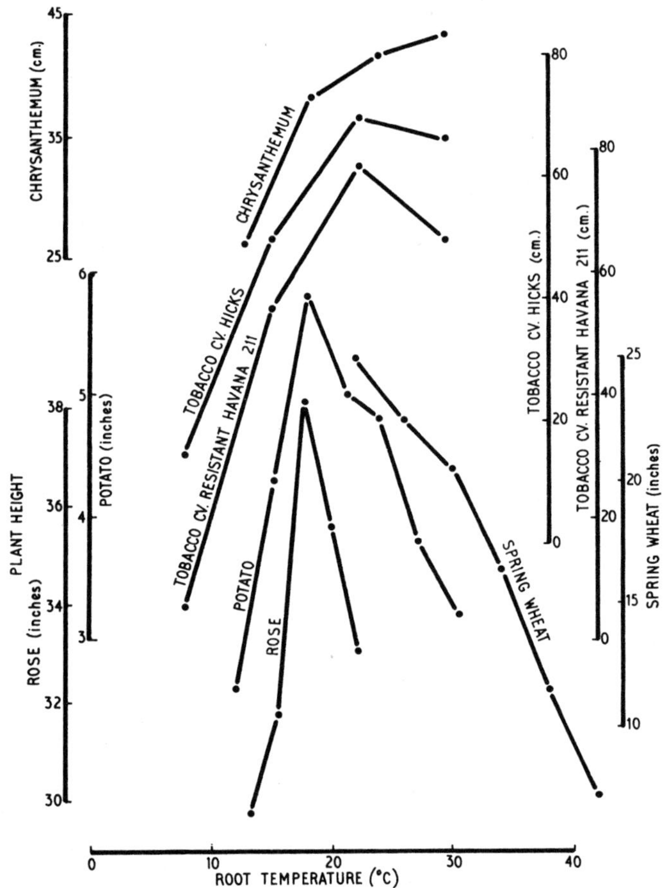

Figure 18: Influence of root temperature on plant height.

cucumbers and tomatoes, Herath and Ormrod (92) for rice, Ashby (8) for *Tilia americana,* Bailey and Jones (11) for blueberries, Arndt (7) and Camp and Walker (45) for cotton, Shanks and Laurie (199) for roses, Richards (187) for potatoes, Parups and Nielsen (171) for tobacco, Wort (244) for spring wheat and Boxall (28) for chrysanthemums.

Figures 17 and 18 suggest that the response curve relating plant height to root temperature may have a similar form to the generalized response curve shown in Figure 8, except for certain species which are exemplified in these data by roses, potatoes and spring wheat and which are characterized by being tallest at relatively low root temperatures. The change in the height of such plants with unit change in root temperature below the optimum is steeper than above the optimum, whereas plants that are tallest at relatively

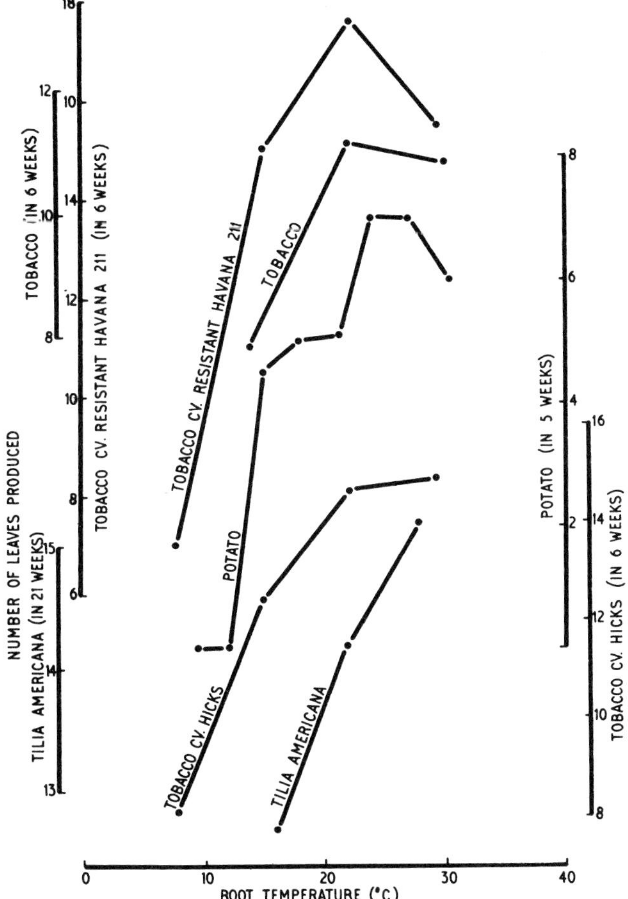

Figure 19: Influence of root temperature on leaf production.

high root temperatures (for example, cotton) exhibit a steeper change above the optimum than below it. While it is clear from Figures 17 and 18 that the heights of plants of different genera may be affected differently by root temperature, cultivars of the same species apparently may not be affected differently; this is indicated by the similar response of two rice cultivars shown in Figure 17.

The height of a vegetative plant can be regarded as the product of the number of nodes and the mean internode length. That leaf production is affected by root temperature is shown in Figure 19 drawn from the data of Ashby (8) for *Tilia americana,* Parups and Nielsen (171) and Parups, Nielsen and Bourget (172) for tobacco and Richards (187) for potatoes, which suggest that there is an optimal root temperature for most rapid leaf production. The data are insufficient to indicate the extent to which the responses of genera, species and cultivars differ. Similarly limited data, which lead to similar conclusions about the response of internode length to root temperature, are illustrated in Figure 20 drawn from the work of Davidson on *Gardenia veitchii* (62), Ashby (8) on *Tilia americana* and Parups and Nielsen (171) Parups, Nielsen and Bourget (172) on tobacco.

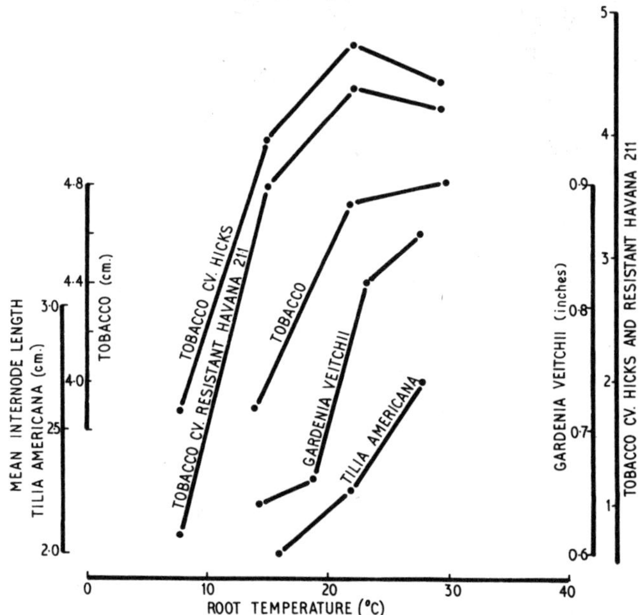

Figure 20: Influence of root temperature on mean internode length.

A little information about the relative importance of leaf production and internode extension in determining plant height can be obtained by comparing the data for the two named cultivars of tobacco in Figures 18, 19 and 20. As root temperature rose from 22 to 29°C so the height of plants of the cultivar

Resistant Havana 211 fell more rapidly than did the height of plants of the cultivar Hicks. This was due almost entirely to the differing influence of root temperature on leaf production, the effect on internode extension differing little between the two cultivars. Above 22°C leaf production by Resistant Havana 211 was adversely affected by an increase in root temperature, whereas leaf production by Hicks plants increased with root temperature between 22 and 29°C.

1.6. Leaf expansion

The influence of root temperature on the total leaf area of a plant is illustrated in Figure 21, drawn from the work of Parups and Nielsen (171) and Parups,

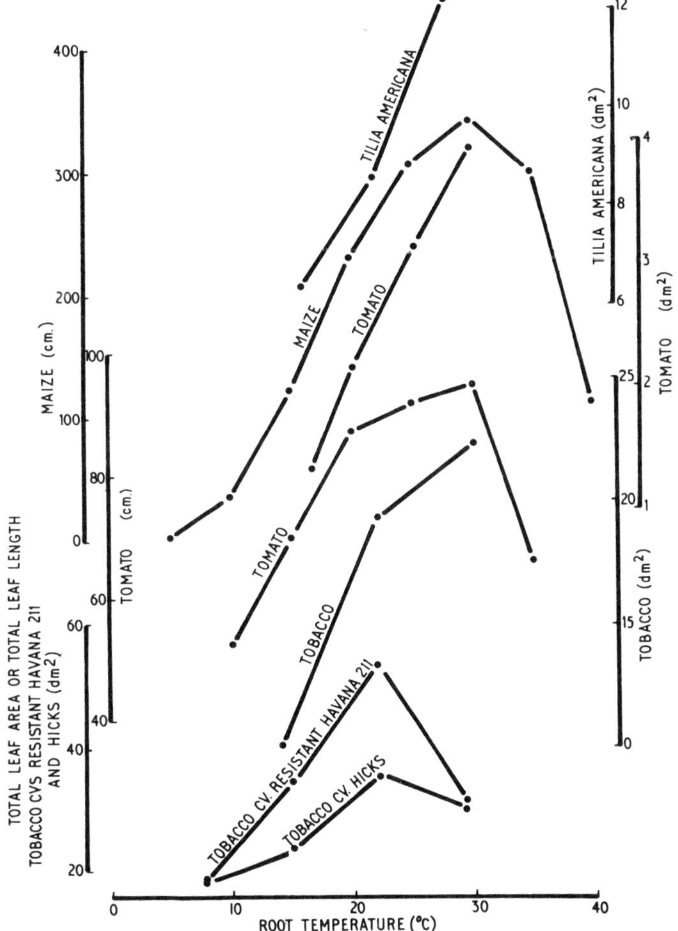

Figure 21: Influence of root temperature on total leaf area (or length).

Nielsen and Bourget (172) on tobacco, Fujishige and Sugiyama (76) and Abd El Rahman, Kuiper and Bierhuizen (1) on tomatoes, Grobbelaar (85) on maize and Ashby (8) on *Tilia americana*. These very limited data suggest that the response curve for leaf expansion may be similar to the generalized response curve of Figure 8.

1.7. Leaf shape and colour

The shape of leaves can be affected by root temperature. Jones, Johnson and Dickson (107) found that tobacco leaves were pointed below 30°C, whereas above 30°C they were rounded. The opposite response was observed by

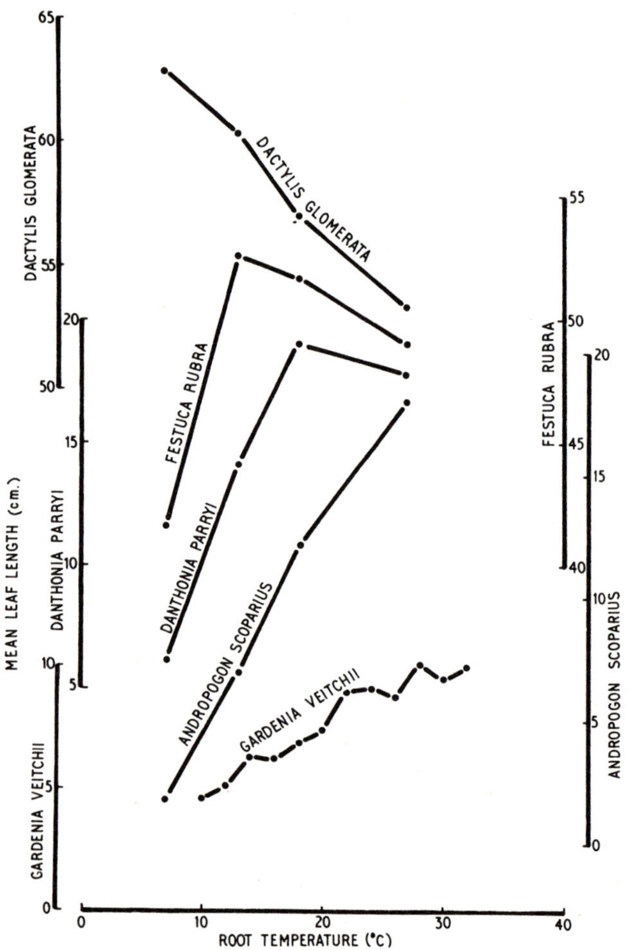

Figure 22: Influence of root temperature on mature leaf length.

Jones, McKinney and Fellows (108) in potato plants, which had rounded leaves at 11°C and pointed leaves at 30°C.

Mature leaf length is affected by root temperature as shown in Figure 22, drawn from the work of Smoliak and Johnston (212) with *Dactylis glomerata, Festuca rubra, Danthonia parryi* and *Andropogon scoparius* and of Jones (106) with *Gardenia veitchii.* The data show that there is an optimal root temperature, which differs between genera, for producing the longest leaves. When the optimum is low the change in leaf length with unit change in root temperature is greater below the optimum than above; when the optimum is high the change in leaf length is greater above the optimum than below.

There are fewer data available defining the influence of root temperature on leaf width. Beauchamp and Lathwell (17) studied the changes in width and length of the first four leaves of maize. At a root temperature of 25°C the lengths of all the leaves were 81-89% of those at 20°C, but the first two leaves had widths that were 107 and 111% of their widths at 20°C, and the third and fourth leaves had widths that were only 97 and 94% Thus, between 20 and 25°C all the leaves became shorter but the first two became wider and the second two narrower. This not only indicates that leaf shape is influenced by root temperature because of differing effects on leaf length and width, but also suggests that there are differences in response between individual leaves.

Root temperature can affect leaf colour. The cotyledons of *Gossypium hirsutum* were a lighter green the higher the soil temperature was between 12 and 24°C, according to Nelson (154). Fujishige and Sugiyama (76) found that tomato leaves were lighter green between 20 and 30°C than at lower or higher root temperatures. Cucumber leaves were lightest in colour between 25 and 35°C.

A different colour phenomenon was observed in *Gardenia veitchii* by Davidson (62). At a root temperature below 19-23°C the leaves were chlorotic, and the lower the temperature the greater was the chlorosis. Jones (106) also reported that at root temperatures above 20°C there was no leaf chlorosis of *Gardenia veitchii* but that there was below 20°C.

1.8. Root extension

The influence of root temperature on root extension is illustrated in Figure 23 drawn from the data of Cannon (47, 48) on *Covillea tridentata,* Adams (2) on white pine, White (238) on the excised roots of tomato, Woodroof (243) on pecans, Nagai and Matsushita (152) on rice, Anderson and Kemper (4) on maize, Galligar (78) on the excised roots of cotton, peas and sunflowers and Barney (13) on loblolly pine. The data of 47, 48, 238 and 13 are for the extension growth of individual roots; those of 243 show the total length of ten roots after four days; those of 78 the total length of twenty roots after fourteen days; and those of 2, 152 and 4 the total length of all roots in the root systems. The similarity to the generalized response curve of Figure 8 is

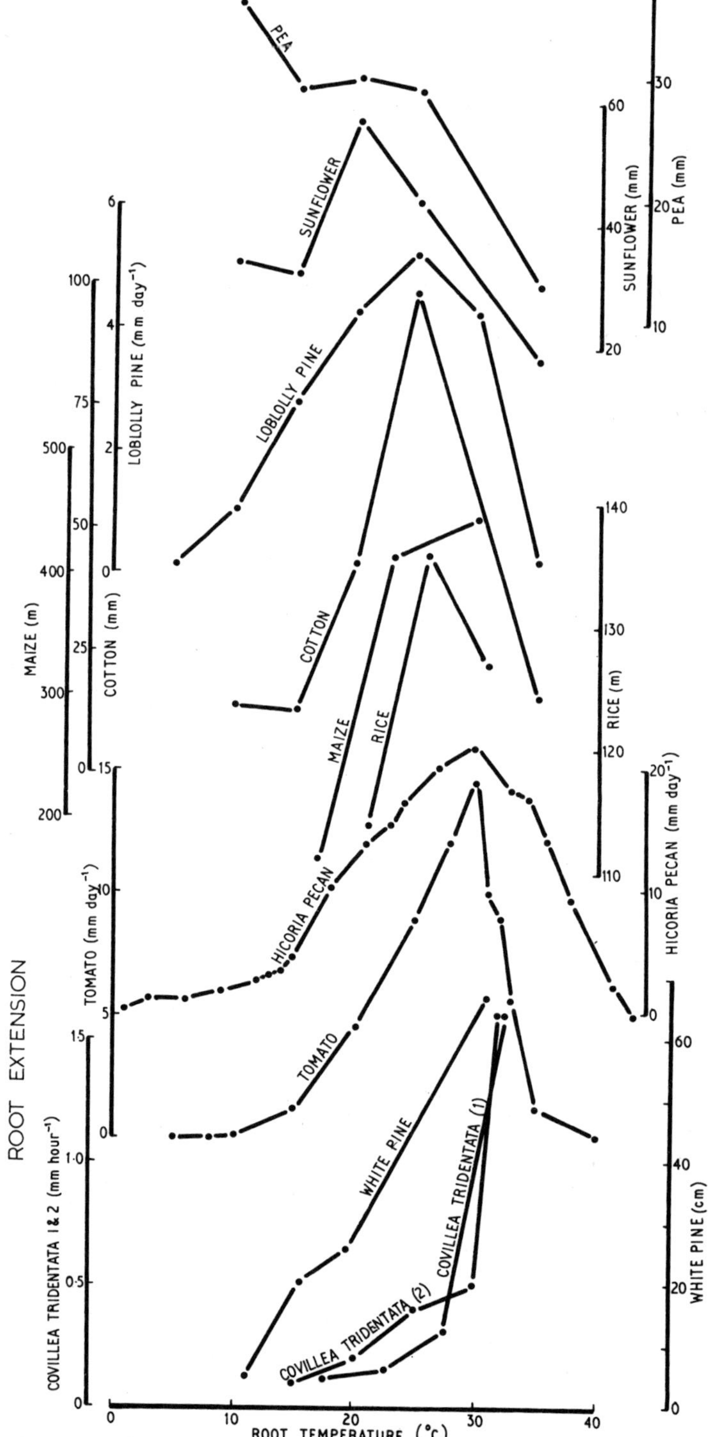

Figure 23: Influence of root temperature on root extension.

again apparent, particularly in the well-defined response of pecan. Differences in response between genera are also evident, an extreme comparison being between pea and *Covillea tridentata*. As the root temperature increases over the range 10-30°C, root extension by pea plants is decreased and that by *Covillea tridentata* is increased. That cultivars within the same species also differ in their response can be seen in Figure 24, drawn from the work of Herath and Ormrod (92) on rice.

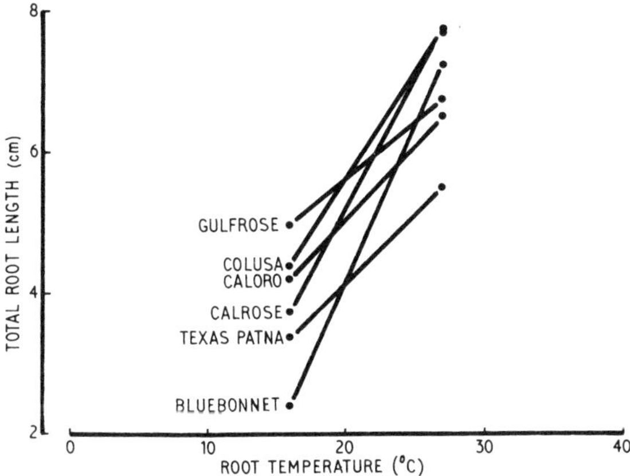

Figure 24: Influence of root temperature on root extension by six rice cultivars.

1.9. Root branching

There appears to be little quantitative information about the influence of root temperature on root branching. Figure 25, illustrating this influence, has been drawn from the work of Galligar (78) with excised roots of maize, sunflower and cotton and Adams (2) with white pine. These data suggest that the effect on root branching is similar to the generalized response shown in Figure 8, and that there are differences in response between genera.

In addition to the quantitative information available several comments have been made about root branching. Garwood (79) observed that the roots of perennial ryegrass were less branched at root temperatures of 7, 8 and 13°C than at 29°C; Nelson and Tukey (156) that the roots of the apple rootstocks M.1, M.2, M.7 and M.9 were less branched at 7 and 13°C than at 19 and 25°C; Roberts and Kenworthy (190) that strawberry roots were less branched at 7 and 13°C than at 18 and 24°C; Nielsen and Cunningham (158) that the roots of *Lolium multiflorum* were less branched at 11°C than at 19.5 and 28°C; Shanks and Laurie (199) that rose roots were less branched at 11°C than at 22°C; Stuckey (216) that the roots of colonial bent grass were less branched at 10°C than at 16°C; and Erickson and Smith (71) that guayule roots were less branched at 16°C than at 21, 27 and 32°C.

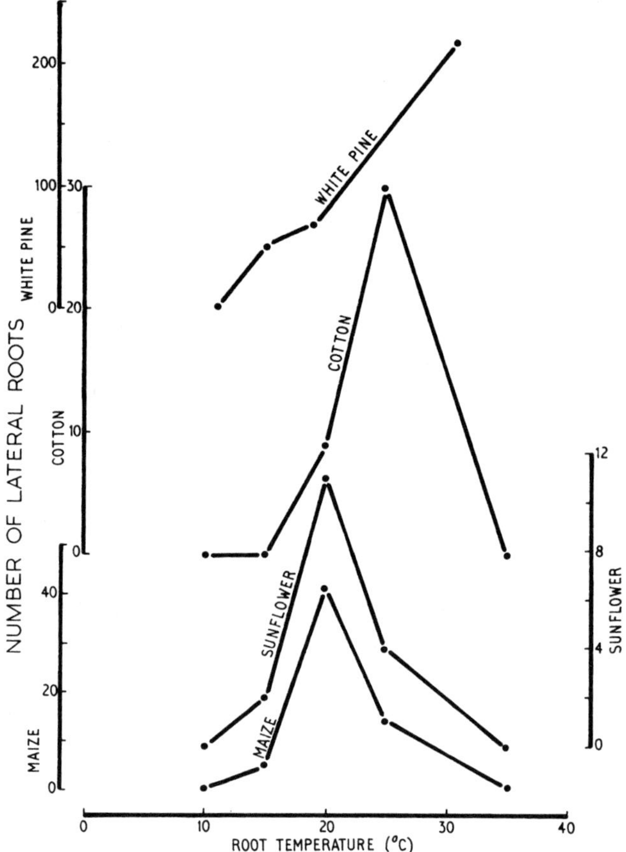

Figure 25: Influence of root temperature on root branching.

1.10. Root thickness

There are few quantitative data on the relationship between root thickness and root temperature. Those illustrated in Figure 26 are from the work of Garwood (79) on S24 perennial ryegrass, Anderson and Kemper (4) on maize and Shanks and Laurie (199) on roses. In each case the mean diameter of the roots decreased as the root temperature increased.

There are also some comments on root thickness. Hellmers (91) noted that the roots of *Sequoia sempervirens* were thick at a root temperature of 18° and thin at 28°C; Ketellapper (112) that the roots of *Phalaris tuberosa* were thick at 15-20° and thin at 30-35°C; Shanks and Laurie (199) that rose roots were thick at 11° and thin at 22°C; Nielsen and Cunningham (158) that the roots of *Lolium multiflorum* were thick at 11° and thin at 28°C; Nelson and Tukey (156) that the roots of the apple rootstocks M.1, M.2, M.7, M.9 and M.16

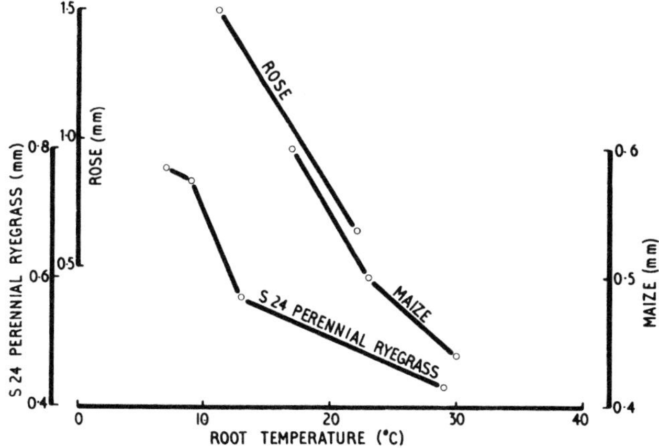

Figure 26: Influence of root temperature on root thickness.

were thick at 7° and thin at 25°C; Roberts and Kenworthy (190) that strawberry roots were thick at 7 and 13° and thin at 24°C; and Stuckey (216) that colonial bent grass roots were thick at 10°, thin at 27° and intermediate in thickness at 16°C.

Thus the general response appears to be that root thickness decreases as root temperature increases within the temperature range used in these experiments.

1.11. Root colour and type

At low root temperatures young roots tend to be white, and at high temperatures brown. Shanks and Laurie (198) reported that rose roots were white at 11° and brown at 22°C; Nielsen and Cunningham (158) that the roots of *Lolium multiflorum* were white at 11° and brown at 28°C; and Ketellapper (112) that the roots of *Phalaris tuberosa* were white at 15-20°C and brown at 30-35°C. Nelson and Tukey (156), working with apple rootstocks, found that at 7°C all roots were white, whereas at 25°C some roots were brown. At 25°C there were more white than brown roots on the rootstocks M.7 and M.16, but more brown than white roots on M.1, M.2 and M.9. Thus the influence of root temperature on root colour differs between cultivars.

Shanks and Laurie (198) noted that rose roots were succulent at 11° and woody at 22°C. Similarly, Nightingale (162) found that at 18° and below apple and peach roots were succulent and snapped easily, but at 24°C and above they were non-succulent and tough. These differences in toughness could be related to the fact that at 24°C the xylem was heavily lignified and there were fibres in the phloem, whereas at 18°C there were no phloem fibres and much less lignification. The differences in succulence could be related to the observation of Shanks and Laurie (199) that rose roots at 11°C contained a greater number of parenchyma cells, which were also larger, than at 22°C. Furthermore, they reported that in rose roots at 11°C there were very few phloem fibres, whereas at 22°C there were many.

The number of root hairs appears to decrease as root temperature rises. Shanks and Laurie (199) found that there were more root hairs on rose roots at 11° than at 22°C; Snow (213) that there were more on wheat roots at 12° than at 24°C, and that at 35°C there were no root hairs; and Gray (84) that there were root hairs on strawberry roots at 29° but not at 38°C.

1.12. Flowering

The temperature of the root system can affect flowering. Proebsting (181) grew strawberry plants at root temperatures of 7, 13, 18, 24, 29 and 32°C and reported that "runners were much more abundant at 24°C than at other temperatures. Flower production was limited to plants grown at lower temperatures." Dickson (65) grew winter wheat at root temperatures from 8 to 36°C, and reported a critical root temperature for flowering between 20 and 24°C. Above 20 to 24°C many tillers were formed and culm elongation was retarded compared to plants at lower temperatures. He reported that the plants "failed to form even embryonic floral organs when held continuously at soil temperatures above 24°C". These findings were supported by Forster (73), who found that root temperatures above 15°C prevented the development of winter wheat plants beyond the rosette stage unless the seeds experienced a period of low temperature soon after imbibition. Tsunoda (225) grew rice plants at root temperatures from 15 to 40°C. At 15 and 20°C the plants had many tillers, a short culm and a short ear; at 25 and 30°C there were few tillers, a long culm and a long ear. Thus, two very different types of plant developed above and below a critical root temperature between 20 and 25°C. At 15°C no initiation of panicles occurred, whereas most panicles were initiated at 20°C. With rising root temperature the number of panicles decreased until at 40°C none was initiated, but at this high temperature the growth of the plants was very abnormal. Inflorescence morphology was also affected; although most panicles were formed at 20°C, most spikelets per panicle were formed at 30°C, at which temperature heading was most rapid and grain yield was greatest.

Differing effects of root temperature on the number of flowers formed are apparent in the work of Allen (3) with *Calendula officinalis* and snapdragons and that of Liebig and Chapman (127) with young orange trees. Figure 27 shows that there were more flowers on *Calendula officinalis* and snapdragons at the higher root temperatures between 11 and 20°C, whereas Liebig and Chapman, although they presented no data, reported that on young orange trees at 14°C there was "considerable blooming, a slight amount of bloom at 22°C and none at 30°C". Davidson (62) found that there was less flower bud drop by *Gardenia grandiflora* and more blooms opened when the root temperature was 23°C than when it was 14, 19 or 28°C.

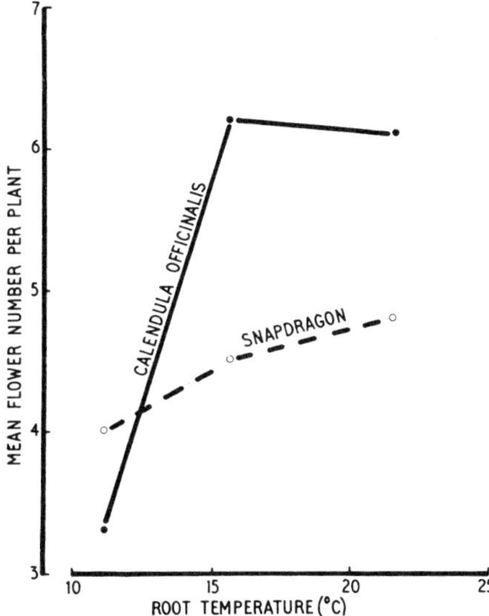

Figure 27: Influence of root temperature on flower number.

1.13. Fruiting

The influence of root temperature on fruiting is illustrated in Figure 28 drawn from the work of Fulton (77) with oats, Mack (135) with barley, Ketcheson (111) with maize and Boxall (25), Bewley (19) and Riethmann (188) with tomatoes. It can be seen that the yield of tomato fruit and the grain yield of oats, barley and maize are affected by root temperature, the data suggesting that the shape of the response curve is similar to the generalized form of Figure 8. Differences in response between genera are apparent: oats have a low optimal root temperature, tomatoes and maize have higher optima and barley is intermediate.

Varying the root temperature for a fairly short period does not appear to affect tomato yields. Boxall (26) raised the root temperature of tomato plants from 16 to 25°C for the month immediately preceding planting in the glasshouse border (i.e. while the plants were still in pots) and found that there was little effect on fruit yield. Similarly, Calvert (44) raised the root temperature of tomato plants from 14 to 18°C for the month immediately following planting and reported that there were "no significant increases in yield of ripe fruit".

Riethmann (188) studied the influence of root temperature on tomato yield in detail and Figure 29 shows the response of fruit number, fruit size (individual fruit fresh weight) and fruit yield (total fruit fresh weight). All three responded similarly to root temperature, with an optimum at about 30-35°C.

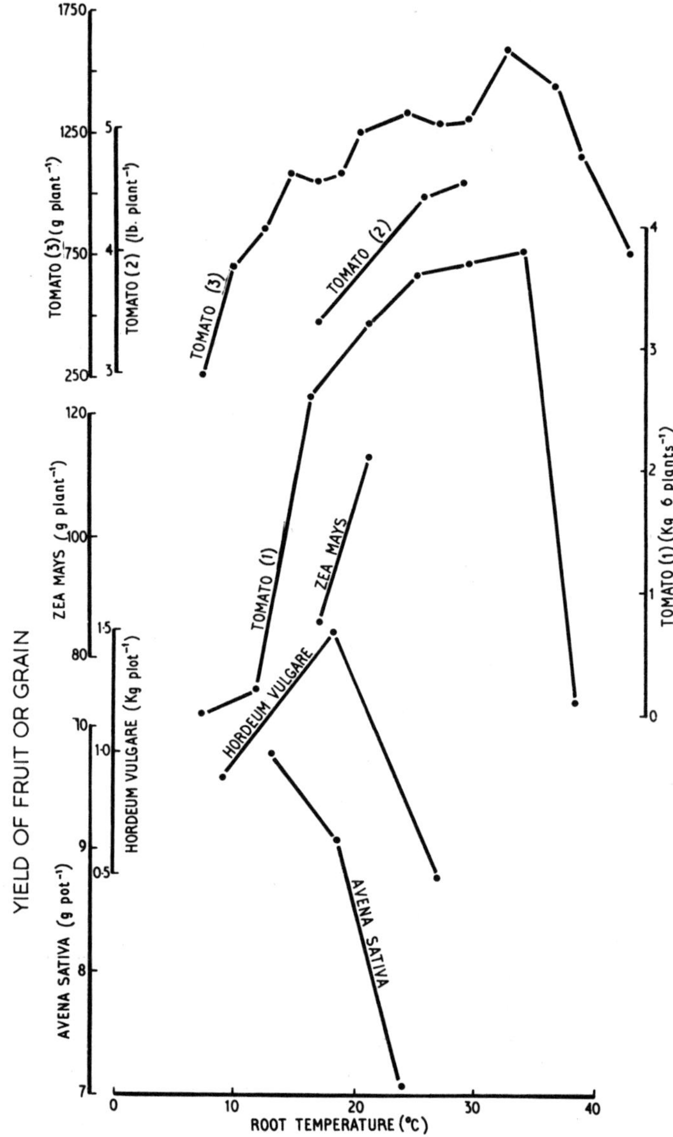

Figure 28: Influence of root temperature on fruiting.

This optimum was compatible with the data obtained by Bewley (19) and corresponded to the optimum of about 34°C determined by Boxall (25), as shown in Figure 28.

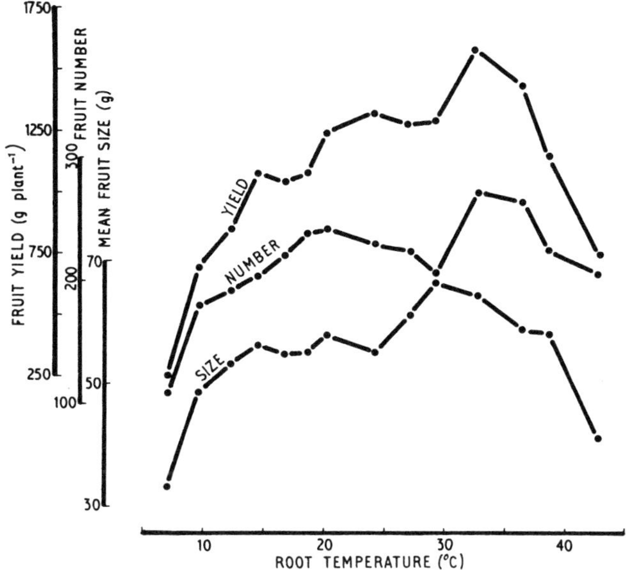

Figure 29: Influence of root temperature on the yield, number and size of tomato fruit.

PART 2

THE INFLUENCE OF ROOT TEMPERATURE ON PLANT PROCESSES

As in Part 1 many of the data in the following five sections have been obtained experimentally in conditions where comparisons have been made between different root temperatures which have been maintained constant throughout the experiments.

2.1. Photosynthesis and respiration

Several workers have reported effects of root temperature on photosynthesis. Vinokur (227) grew lemon plants at root temperatures of 13 and 33°C and measured photosynthesis "by the manometric method in Warburg's apparatus at 25°C on leaf sections of 4 cm^2 and an exposure of 10 minutes". He stated that "the prolonged action of temperature on roots affects the intensity of photosynthesis. When the value at 33°C is taken as 100%, at 13°C it is 56-61%". Similarly, Andreenko and Kerechki (5) observed that net photosynthesis by maize plants was less at a root temperature of 12° than at 25°C. They stated that "the amount of chlorophyll per unit dry weight of leaves proved to be noticeably less at the lower temperature" and concluded that "the chlorophyll content of the leaves is quite clearly positively correlated with the intensity of photosynthesis". Between 10 and 27°C the net photosynthesis of ***Pinus radiata,*** measured by recording the rate of CO_2 consumption with a Heath recorder connected to an infra-red gas analyser, increased with root temperature according to Babalola, Boersma, and Youngberg (10). A similar effect on rooted bean leaves ***(Phaseolus vulgaris)*** was observed by Humphries (97) from dry weight determinations at root temperatures of 12 and 24°C; net assimilation rate was greater at the higher temperature. Grobbelaar (85), using fresh weight determinations, compared the influence of root temperature at 5-degree intervals from 5 to 40°C on net assimilation by young maize plants. Between 15 and 35°C inclusive there was little effect of root temperature, but at lower and higher temperatures net assimilation rate was reduced (Figure 30).

All these reports support the detailed work of Brouwer (37) who maintained eight root temperatures at 5-degree intervals between 5 and 40°C with maize, red kidney beans and peas (Figure 30). The net assimilation rate, obtained from fresh weight determinations, decreased in maize plants below 15°C and in kidney beans below 10°C, although in peas there was no reduction even at 5°C. Above 25°C there was a decrease in peas and above 35°C in kidney beans, but at 40°C there was little adverse effect on maize. Thus, the shape of the response curve of net assimilation rate to root temperature differed between species, but in all the species examined there was a broad optimal temperature band suggesting that, in general, net assimilation rate may be independent of root temperature except at extremes of root temperature.

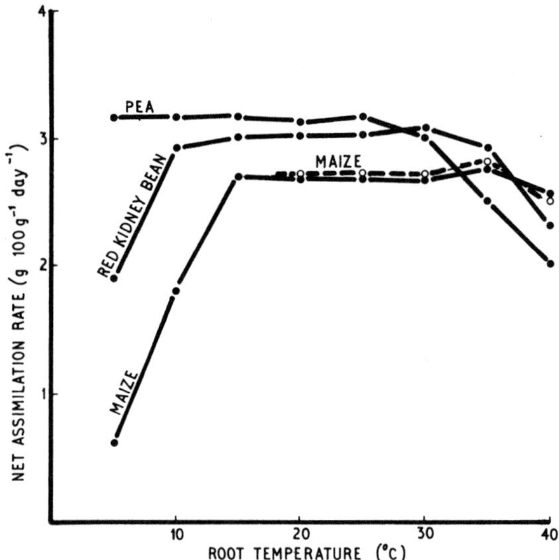

Figure 30: Influence of root temperature on net assimilation rate.

The influence of root temperature on chloroplast structure was studied by Shakhov and Golubkova (197). They grew potato and tomato plants at root temperatures of 8 and 20°C, and reported that at 20°C "in the leaves of the potato the chloroplasts are dense and the stroma contain grana which are arranged in several layers. These were small and closely appressed so as to give the impression of a compact mass. Osmiophilic granules were large and dispersed throughout the chloroplast, their diameter being about half that of the grana. Chloroplasts from plants grown in cold soil were larger, as were their grana, which were loosely disposed in one layer in the stroma, and osmiophilic granules were very small." They also observed that at a root temperature of 20°C the "chloroplasts of tomato are rounded with clearly visible grana and large osmiophilic granules . . . at the lower soil temperature . . . disintegration occurs in the chloroplasts: first, chloroplasts with no clearly distinguished grana; second, those with only osmiophilic granules in the stroma; third, stroma fragments with osmiophilic granules surrounding them. It is therfore chiefly the grana and stroma which disintegrate".

Respiration is also affected by root temperature. Jensen (101) grew maize and tomato plants at root temperatures ranging from 5 to 35°C at 5-degree intervals and determined the respiration of the roots. Respiration increased with temperature, but the effect of a 10°C rise became less as the root temperature rose. Korovin and Barskaya (117) determined the respiration of roots of wheat, maize and potato plants at root temperatures of 10 and 20°C. They reported that "at the higher soil temperature, intensity of root respiration was about the same in wheat, potato and maize. However, upon lowering the temperature the intensity of root respiration in maize decreases more markedly than in wheat and potatoes". They also found that a change in respiration intensity as a result of root temperature change was observable within an hour. For shoots no effect of root temperature between 10 and 27°C on respiration of ***Pinus radiata*** was observed by Babalola, Boersma and Youngberg (10), although according to Andreenko and Kerechki (5) the maize shoot has a faster respiration rate at a root temperature of 25° than at 12°C.

2.2. Water absorption

The influence of root temperature on water absorption is shown in Figure 31 drawn from the work of Takeshima (220) on rice, Unger and Danielson (226) on beans *(P. vulgaris)* and Brouwer and van Vliet (39) on peas. Their data indicate that there is an optimal root temperature for water absorption, which differs between species, when root temperatures are maintained constant. Kuiper (121) also studied the effect of a change in root temperature on water absorption by beans. He reported that a "change in root temperature induced an immediate change in water uptake giving rise to a new, constant level after about 15 to 30 minutes. Then another change in water uptake occurred, continuing until a new, constant level was reached after 30 to 60 minutes." He attributed the first of these changes to the influence of temperature on the

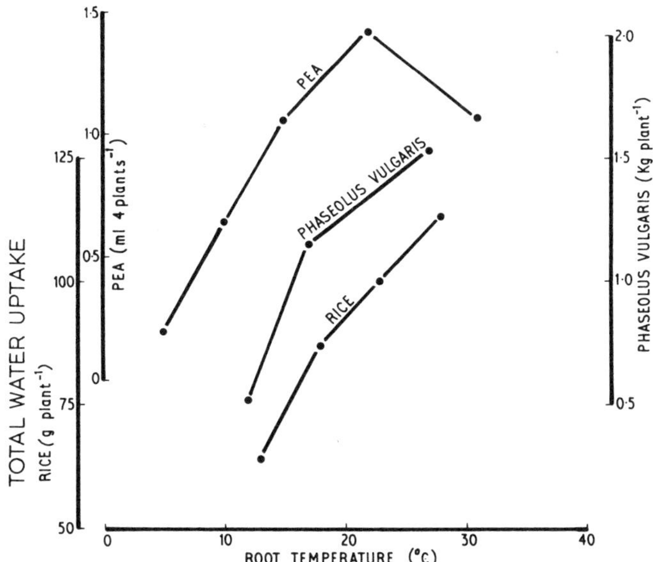

Figure 31: Influence of root temperature on water uptake.

viscosity of the water and the second to an effect on the water permeability of the cytoplasmic membranes. He grew plants at root temperatures of 17 and 24°C, and determined water uptake, after both the above changes in uptake had occurred, when the root temperature was changed to values between 5 and 32°C. Plants that had previously been grown at 24°C showed a pronounced, positive relation between water absorption and the new temperature over the whole range from 5 to 32°C. These findings are similar to those of Tagawa (218), who grew bean seedlings at a root temperature of 20°C and then transferred them to 0, 5, 10, 15, 20, 25 and 30°C. After the rate of water absorption had settled down to a constant value following the transfer it was found that the higher the temperature the greater was the rate of water absorption; relative to absorption at 20°C (taken as 100%) the range was from 57% at 0°C to 130% at 30°C. However, Kuiper (121) found that when the plants had previously been grown at 17°C there was a pronounced positive relationship between absorption and the new root temperature between 5 and 17°C only; from 17 to 32°C the effect of the root temperature on water uptake was much smaller. These data suggest that the previous root temperature history of a plant affects its response to a subsequent root temperature. In both groups of plants, however, reducing the root temperature reduced water absorption. This conclusion is supported by Duncan and Cooke (67), who found that lowering the root temperature of sugar cane from 28° to 21, 15 and 10°C caused progressively greater decreases in water absorption, and by Raleigh (183), who found that decreasing the root temperature of muskmelon plants from 32° to 27, 18 and 10°C had a similar effect on water absorption.

Kramer (120) determined the water absorption of *Gossypium hirsutum*, watermelon, kale, *Pinus taeda, P. caribaea, P. resinosa* and *P. strobus* and observed that lowering the root temperature reduced water absorption, but that the response differed with species, for example, a reduction from 25 to 10°C reduced absorption to 20% with watermelon but to only 75% in kale.

Böhning and Lusanandana (21) drew attention to the fact that "under natural conditions, the change in temperature varies not only in respect of the level to which the temperature falls but also in the rate at which it occurs". Working with tomatoes they gradually decreased root temperature from 25 to 5°C over 13 days, rapidly decreased it over the same range, or left it unchanged at 25°C. Water absorption by the tomatoes slowly increased to a maximum at 18°C as the temperature gradually decreased, but below 18°C the rate of absorption declined with a fall in root temperature. Gradually lowering the temperature gradually reduced water absorption, but an abrupt change in temperature resulted in an abrupt decrease in absorption to a value similar to that of the slowly cooled plants. Ehrler (68) found that, although a rapid reduction in root temperature from 24 to 5°C reduced the water absorption of lucerne by 70% during the first 24-hour period of low temperature, there was some recovery, and that during the second 24-hour period the reduction was only 50%.

2.3. Water movement and transpiration

If the end of a root is cut off and the rate of exudation from the cut surface is measured, an estimate of the rate of movement of water through the root is obtained. O'Leary (164) observed that the exudation from the cut surface of white spruce and loblolly pine roots was greater at 34° than at 24°C. O'Leary (165) also found that with cotton and castor bean roots the exudation was greater at 35° than at 25°C. On the other hand, in bean and sunflower roots it was less at 35° than at 25°C, and in maize roots temperatures between 25° and 35°C had little effect. Thus, it seems probable that there is an optimal root temperature (or range of temperature) for water movement along roots, which differs with species.

If a plant has its shoot cut off a little above the soil surface, measuring the rate of exudation from the cut surface of the remaining stem will give an estimate of the movement of water through the roots and the stem up to the cut surface. Grossenbacher (86) observed that, when a sunflower plant was decapitated just below the cotyledons, the rate of exudation from the cut surface was greater at a root temperature of 30° than at 15°C. Kramer (119) found that when sunflower and tomato plants were decapitated at the first node the exudation from the cut surface was greatest at a root temperature of 25°C, and exudation ceased at 12°C in tomatoes and 2.5°C in sunflowers. These findings, as well as suggesting that there is an optimal root temperature for water movement up to the first node, also suggest that the shape of the response curve may differ between species having the same optimum.

The influence of root temperature on transpiration is shown in Figures 32 and 33, drawn from the data of Brouwer and van Vliet (39) on peas, Bialoglowski (20) on rooted, leafy lemon cuttings, Kozlowski (118) on *Pinus strobus* and *P. taeda*, Brouwer (35) on peas and *Phaseolus* beans, Barney (13) on loblolly pine, Franco (74) on coffee, Grobbelaar (85) on maize, Clements and Martin (54), Rajan (182) and Whitfield (239) on sunflowers, Cox and Boersma (55) on white clover, Ongun and Wallis (167) on oranges and Takeshima (220) on rice. The root temperature at which transpiration is greatest differed between species, as did the width of the optimal temperature band, this being very broad in sunflowers.

The work of Gray (84) with strawberries, which is illustrated in Figure 34 (right), suggests that, although cultivars may show some differences in response, the basic response of the species is dominant and that, regardless of the cultivar, an increase in root temperature from 29 to 39°C reduced transpiration. Similarly, the work of Babalola, Boersma and Youngberg (10), illustrated in Figure 34 (left), suggests that, although the availability of water affected the response of transpiration to root temperature, it did not mask the basic response of the species, namely, an increase in the transpiration of *Pinus radiata* as the root temperature rose from 10 to 27°C.

In studies undertaken by Bialoglowski (20) the influence of root temperature on transpiration was manifest only during the day. He attributed this to the fact that transpiration during the night by the rooted, leafy lemon

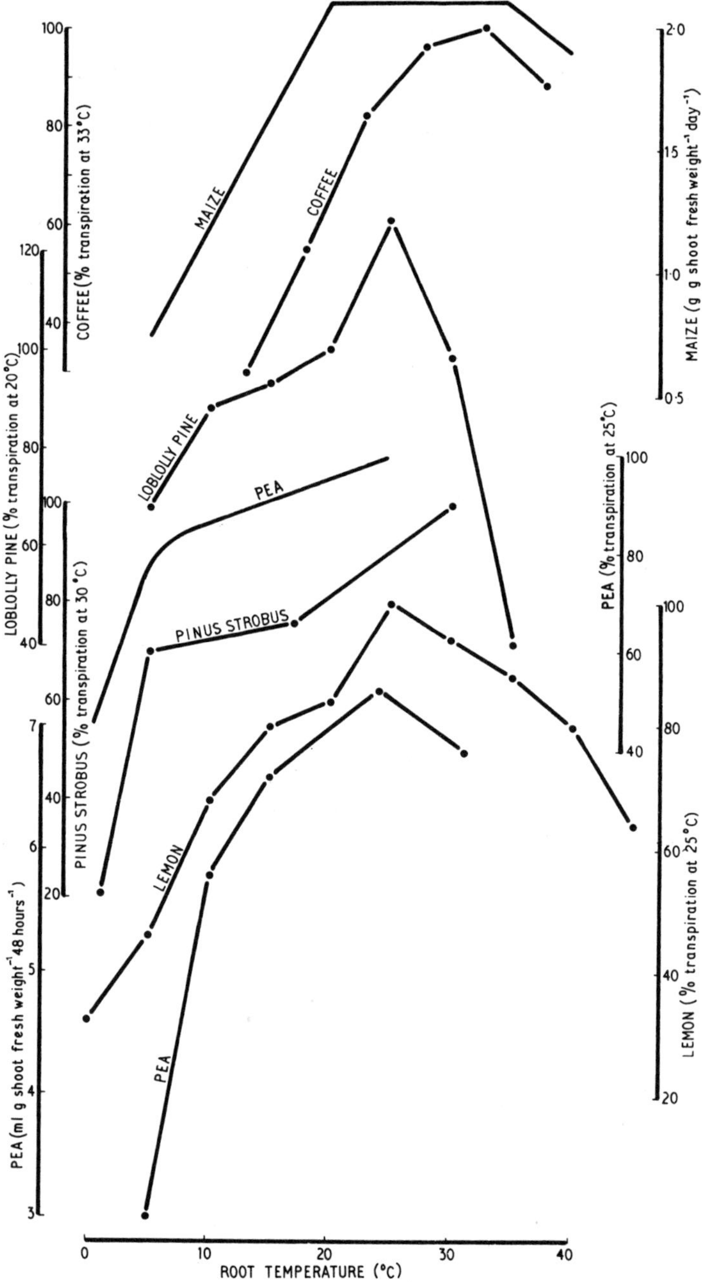

Figure 32: Influence of root temperature on transpiration.

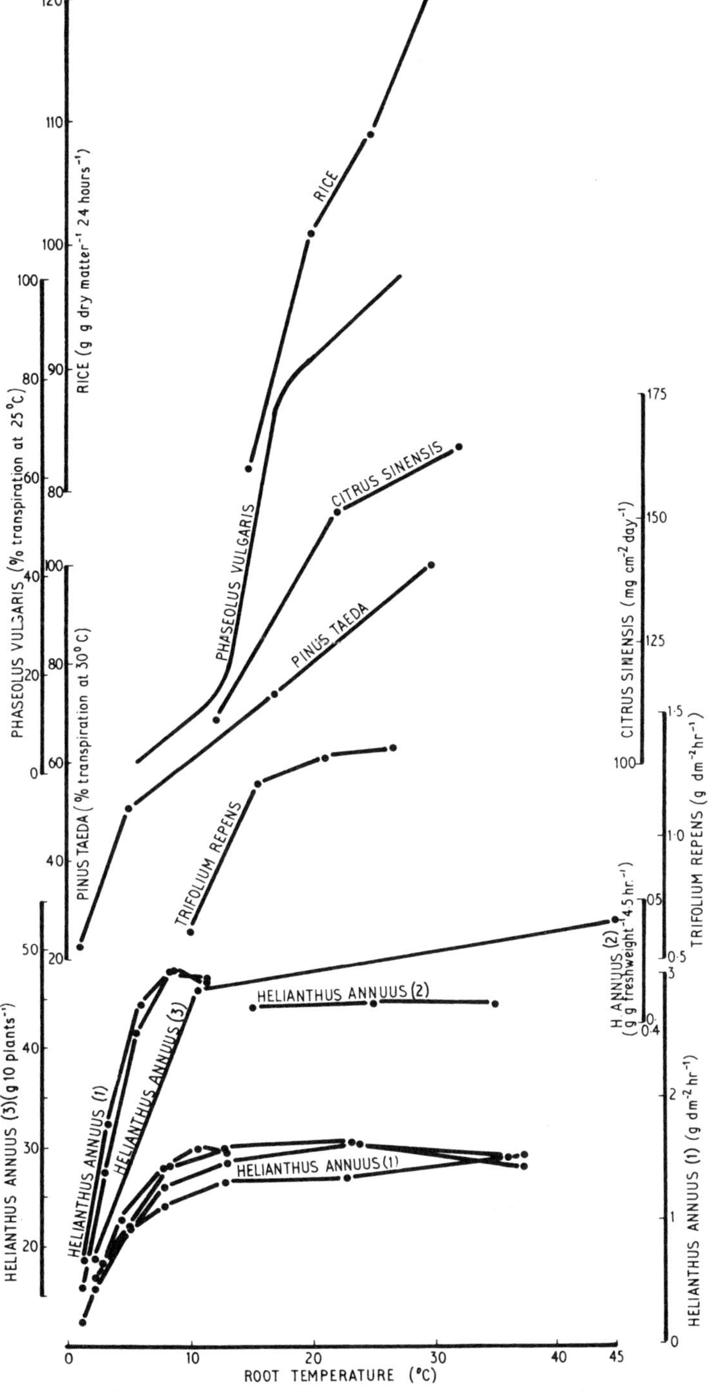

Figure 33: Influence of root temperature on transpiration.

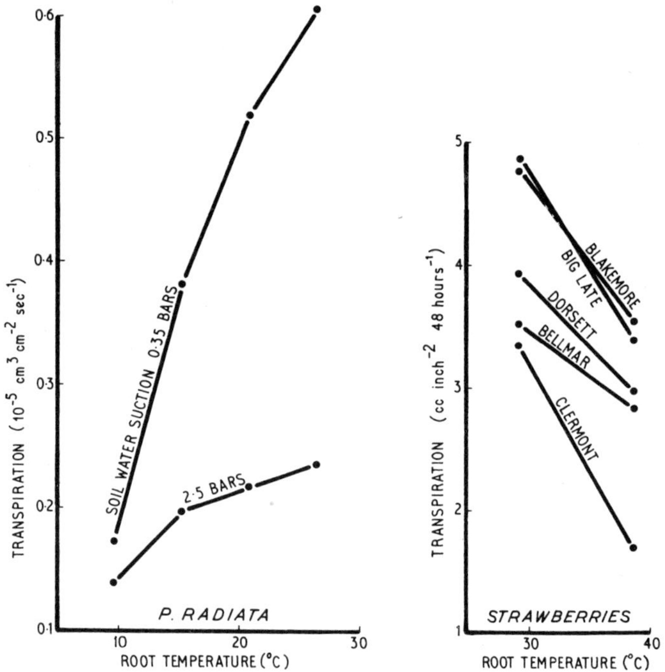

Figure 34: Influence of soil water suction (left) and cultivar (right) on response of transpiration to root temperature.

cuttings with which he worked was less than their transpiration during the day at a root temperature of about 0°C.

In determining the influence of root temperature on transpiration the data of Barney (13) should be noted, if plants are to be grown at one root temperature which is subsequently changed. When loblolly pine seedlings were grown at 30°C and the root temperature was increased to 35°C the transpiration rate was still falling five days after the temperature change. When the root temperature was lowered to 10°C, transpiration decreased for two days and then increased, and after five days it was still rising and had not stabilized. The determination of the time required to achieve stabilization of the response that is being studied is thus important when a change of root temperature is involved.

That a change in root temperature affects the water balance of plants was shown by the observation of von Sachs (191), that when there is a rise in root temperature many species exude droplets of water at the edges of their leaves, and that when the root temperature falls they lose turgidity. The effect differs with different species. Böhning and Lusanandana (21) found that when root temperature fell from 25 to 15°C neither bean, sunflower nor tomato plants wilted, but at 10°C bean plants did wilt, and at 5°C both the sunflower and tomato plants also wilted. An adjustment to the new temperature took place; nevertheless the sunflower and tomato plants wilted severely for the first day and then recovered turgidity without visible injury, but part of the bean plants' foliage was killed and the wilting disappeared very slowly. Unger and Danielson (226) observed that when the root temperature of bean plants fell from 25 to 18°C there was slight wilting, but turgor was regained rapidly; if, however, the temperature fell to 10°C wilting occurred within half-an-hour and turgidity was not regained until the second day. Burkholder (43) also reported slight, temporary wilting of bean plants when the root temperature fell to 18°C. When the root temperature of muskmelon plants fell from 32 to 18°C Raleigh (183) observed slight wilting, but when it fell to 10°C severe wilting developed rapidly. When the root temperature of cotton fell from 27 to 15°C, Guinn and Hunter (87) also observed slight wilting, which disappeared after one day, but at 10°C there was prolonged wilting. Nelson (154) similarly reported that when root temperature fell from 28 to 18°C there was no wilting of cotton, whereas at 12°C there was severe wilting within an hour. Clements and Martin (54) observed that sunflower plants wilted when the root temperature fell to 4°C but recovered turgidity rapidly if it rose again. Post and Mastalerz (178) found that when the temperature of the soil in which poinsettias were growing fell to about 7°C the plants wilted severely for two days, but recovered turgidity on the third day. The leaves of orange trees wilted when the roots were colder than 17°C (North and Wallace, 163), and cucumber plants wilted when the root temperature fell from 29 to 16°C (Schroeder, 193). There was some wilting of the leaves of blueberry bushes when the roots cooled below 18°C (Bailey and Jones, 11).

Finally, a further note of caution about experimental technique is called for; Kramer (120) demonstrated that the *rate* at which root temperature fell affected the wilting response. Rapid cooling induced more severe wilting than slow cooling.

2.4. The mineral content of plant tissue

The influence of root temperature on the nitrogen content of the whole plant is shown in Figure 35, derived from the work of Nielsen, Halstead, McLean, Holmes and Bourget (160) on oats, of Shtrausberg (201) on tomatoes and cabbages, Franco (74) on coffee, Jones and Tisdale (103) on soybeans, Grobbelaar (85) on maize, Heinricks and Nielsen (90) on lucerne and of Carlson (49) on *Malus sylvestris*.

The influence of root temperature on the phosphorous content is shown in Figure 36, drawn from the work of Carlson (49) on *Malus sylvestris,* Parups, Nielsen and Bourget (172) on tobacco, Wallace, Romney, Hale and Hoover (232) on *Phaseolus* beans, maize, soybeans and *Gossypium hirsutum,* Knoll, Brady and Lathwell (115) and Grobbelaar (85) on maize, Hori, Arai, Hosoya and Oyamada (95), Locascio and Warren (130), Lingle and Davis (128), Martin and Wilcox (143) and Wilcox, Martin and Langston (241) on

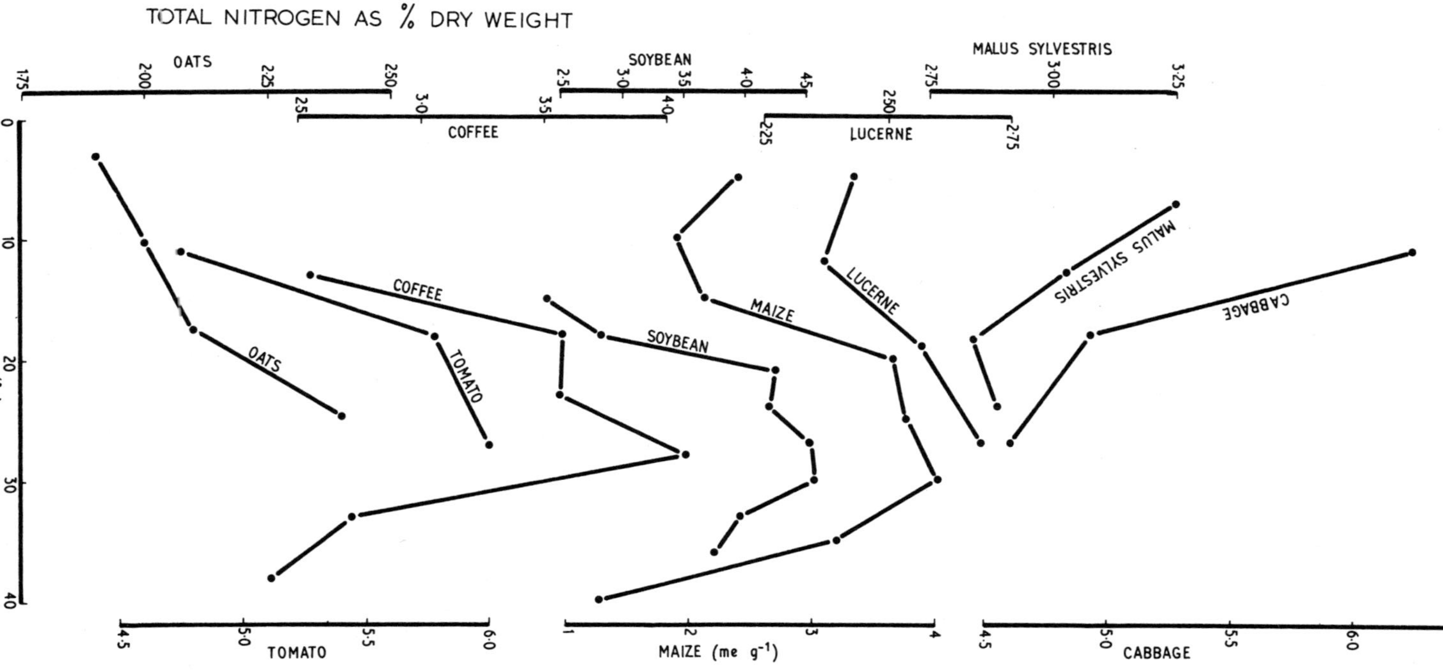

Figure 35: Influence of root temperature on nitrogen content of plant tissue.

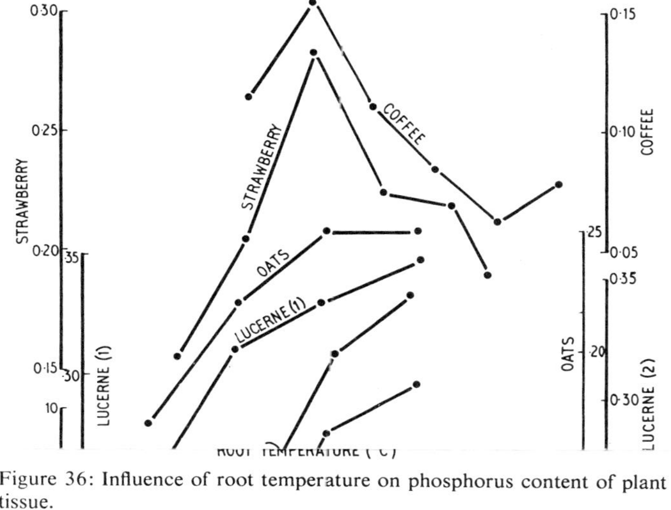

Figure 36: Influence of root temperature on phosphorus content of plant tissue.

[facing p. 40

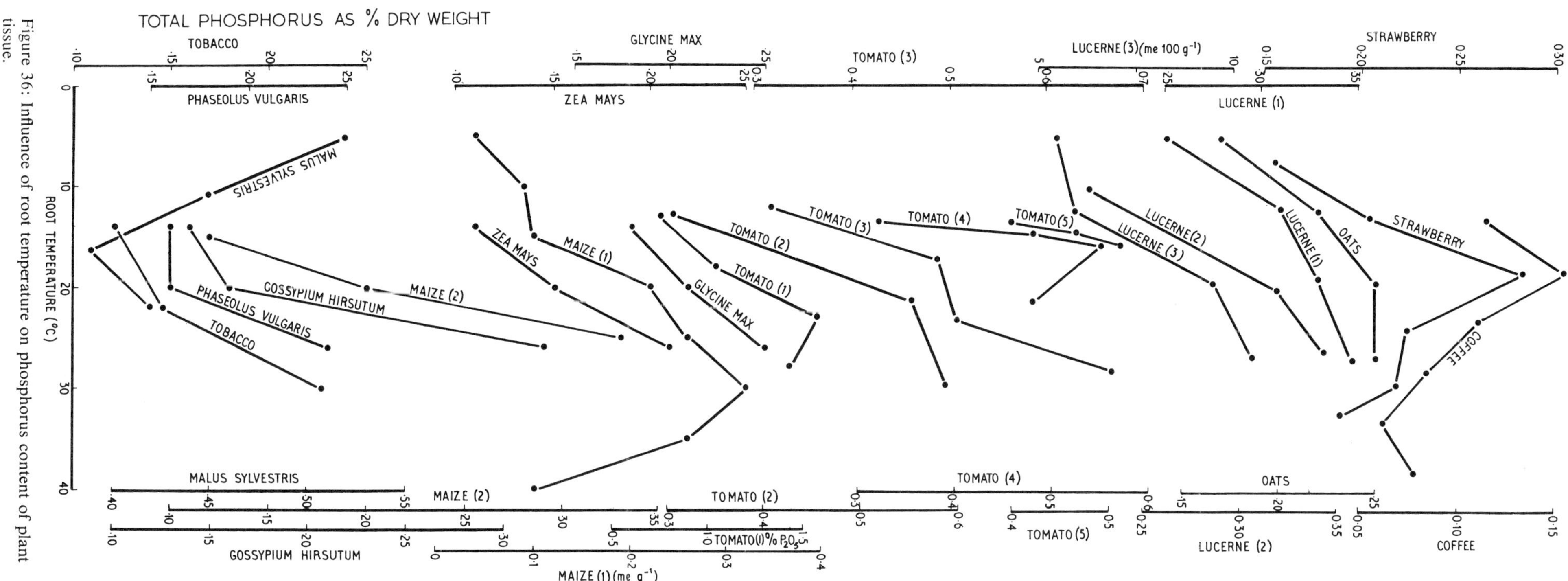

Figure 36: Influence of root temperature on phosphorus content of plant tissue.

[facing p. 40

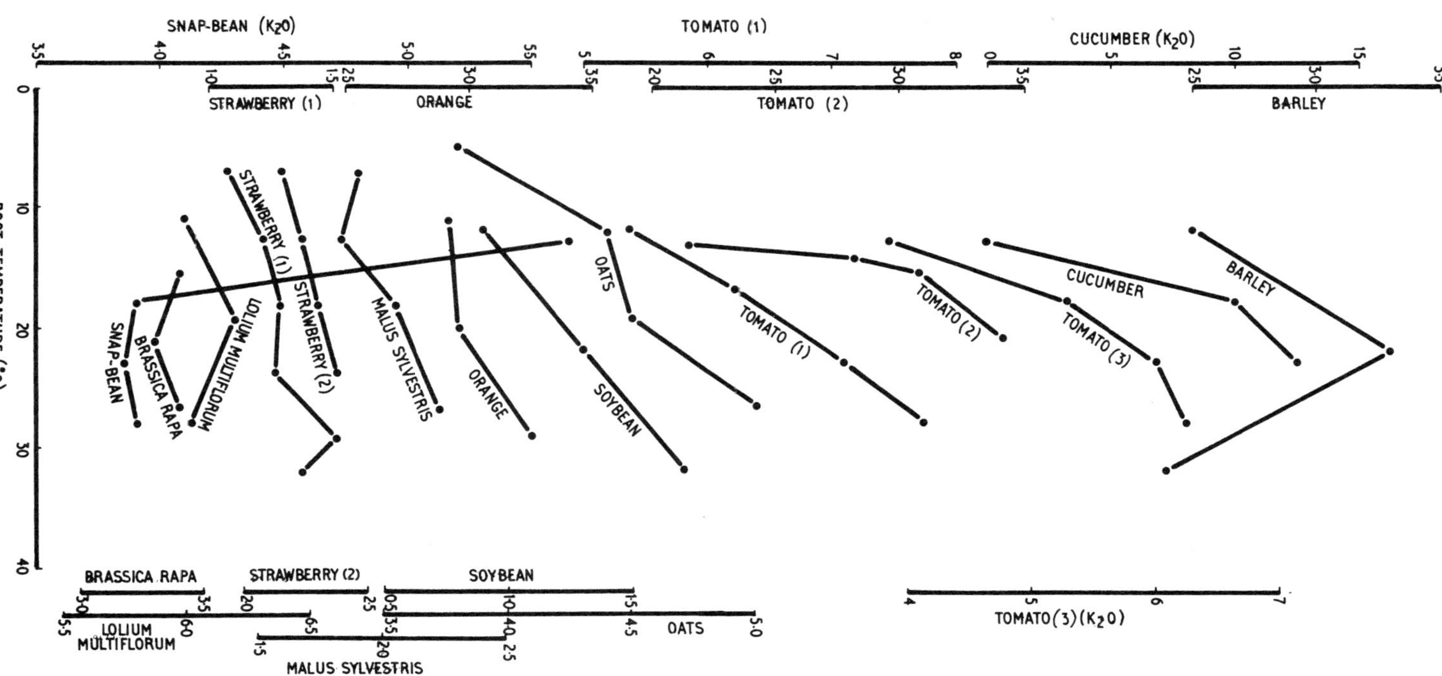

Figure 37: Influence of root temperature on potassium content of plant tissue.

tomatoes, Nielsen, Halstead, MacLean, Holmes and Bourget (161), Levesque and Ketcheson (124) and Heinricks and Nielsen (90) on lucerne, Nielsen, Halstead, MacLean, Holmes and Bourget (160) on oats, Proebsting (181) on strawberries and Franco (74) on coffee.

Lastly, the influence of root temperature on the potassium content is shown in Figure 37, derived from the work of Hori, Arai, Hosoya and Oyamada (95) on snap beans, Army and Miller (6) on turnips, Nielsen and Cunningham (158) on ***Lolium multiflorum,*** Proebsting (181) and Roberts and Kenworthy (190) on strawberries, Carlson (49) on *Malus sylvestris,* Ongun and Wallace (166) on orange, Wallace (231) on soybeans, Nielsen, Halstead, MacLean, Holmes and Bourget (160) on oats, Lingle and Davis (128), Martin and Wilcox (143), and Hori, Arai, Hosoya and Oyamada (95) on tomatoes and cucumbers and Wallace (231) on barley.

These data suggest that there exists a general relationship, perhaps applicable to all species, between root temperature and the percentage mineral content of plant tissue. This generalization is illustrated in Figure 38. Curves A, B, C and D indicate four representative species. In species A the percentage mineral content decreases with increasing root temperature over most of the range between 0 and 40°C, whereas, in species D, there is an optimal root temperature at which mineral concentration is greatest. The response curve moves between these two extremes as shown. The high concentration at low temperature decreases along the line VW, and a high temperature peak develops along the line YZ. These two movements create a trough of low concentration which moves along the line XW.

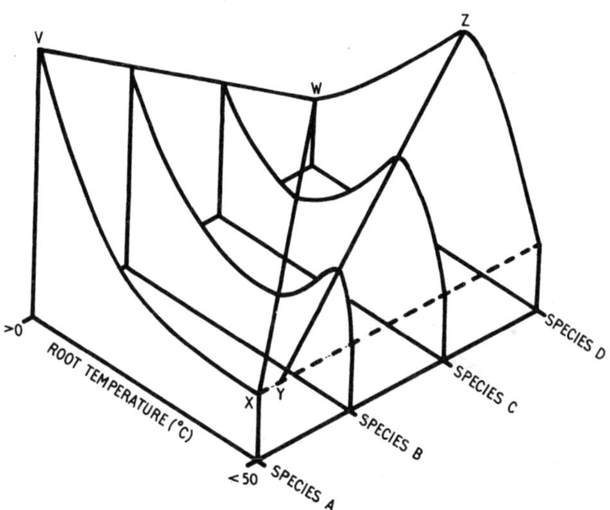

Figure 38: General relation for all species between root temperature and % mineral content of plant tissue.

Specific examples for the nitrogen content are provided in Figure 35 by cabbage (which approximates to species A), *Malus sylvestris* (species B), maize (species C) and oats (species D). For the phosphorus content examples in Figure 36 are *Malus sylvestris* (species B) and maize (species D). For the potassium content examples in Figure 37 are snap beans (species A) and tomatoes and barley (species D). Nevertheless, the data suggest that the majority of species approximate to species D as far as nitrogen, phosphorus and potassium content are concerned. This implies that, for most plants, the effect of root temperature on mineral uptake is proportionately greater than the effect on growth.

It is of interest that the generalization in Figure 38 could be achieved without adjustment of the generalized response curve of Figure 8 if, for example, the relationships between root temperature and (a) mineral uptake and (b) growth have, for a given species, the common from shown in Figure 8, but different optimal temperatures and a slightly different temperature range. Because the mineral content of plant tissue is directly proportional to the rate of mineral uptake and inversely proportional to the rate of growth, the situation illustrated in Figure 38 would arise.

PART 3

NATURAL CONDITIONS OF ROOT TEMPERATURE

So far, this review has been largely factual. A sufficent selection of data about growth phenomena and plant processes has been presented to give an indication of the influence of root temperature. There has been little speculation, except in the presentation of Figures 8 and 38 which summarize the general influence of root temperature. It is reasonable to speculate that when the response to root temperature is mediated primarily through one influence, the response is of the form shown in Figure 8. On the other hand, when the response is mediated by two influences (for example, the percentage nitrogen content of plant tissue is influenced by both the growth of the tissue and its receipt of nitrogen) then the response to root temperature could be of the form shown in Figure 38, although many forms of this curve are to be expected depending on the detailed nature of the response to the two influences. This is because of the interaction of two influences whose response curves, while having the general form shown in Figure 8, have slightly dissimilar optimal temperatures and temperature ranges. If, therefore, the underlying basic response to root temperature of most growth phenomena and plant processes is of the form shown in Figure 8, one of the most fundamental unanswered questions concerning the influence of root temperature is to explain why the generalized response curve has the form shown in Figure 8.

It is a reasonable hypothesis for disproof that root temperature exerts its influence on plant growth by affecting one or more processes occurring in the roots. These processes can be grouped under the three headings of:

1. Synthesis.
2. Absorption and translocation within the root.
3. Structural changes.

Under the first heading of "Synthesis" one theory that could be advanced to account for the sigmoid rise to the optimum of the response curve in Figure 8 is the influence of temperature on the synthesis of growth-regulating substances by the roots. Under the second heading it has already been shown that the absorption of water increases with temperature up to an optimum; therefore, under "Absorption" an alternative theory could be advanced based on water absorption, while another possibility is nutrient absorption. Under the third heading "Structural changes" the decline in the response curve above the optimum could well be attributed to an increase in suberization of the roots.

If, however, root temperature does not exert its influence on plant growth by affecting processes occurring in the roots, another possibility to consider is its influence on translocation from the root to the shoot. There are several possible ways in which this could take place. Thus, if growth were being controlled by the amount of a growth-regulating substance in the shoot, then the influence of root temperature on its translocation from the root could be a limiting factor. It would also be quite reasonable to assume that root temperature influences the translocation of some other substance as the limiting factor.

If the hypothesis that root temperature exerts its influence by affecting translocation from root to shoot is unacceptable, then it may be that translocation within the shoot is being affected.

It is apparent from the earlier sections, in which evidence was reviewed, that it would be possible to obtain many good correlations between the influence of root temperature on growth and its influence on various plant processes, each of which would require testing to show the existence of a causal relationship. This would sometimes be a long and difficult task. Some elimination of possibilities might be achieved, however, if, after having determined the growth response curve to root temperature, selected points on the curve were redetermined with one other factor of the environment held at a very different level. Determining which factors interacted with root temperature, and considering the form of the interactions, might indicate the mechanisms involved.

It must be admitted that the statement of Brouwer (38) made in 1964 still remains substantially correct, that "though the effect of root temperature on the growth of various crops has been investigated relatively often, it is as yet impossible to gain a clear impression of the causal relations which are of importance". In all the experimental work presented so far, however, the effects of constant root temperatures have been examined, but under natural conditions plant roots rarely experience constant root temperatures. It might, therefore, be helpful before proceeding further to consider naturally occurring soil temperatures.

3.1. Soil temperature

The surface of very dark soil in India can reach as high a temperature as 75°C (185) and bare soil in the central Rocky Mountains as high as 71°C (16). The surface of sandhills in Nebraska can reach a temperature of 68°C (177) and in Minnesota 56°C (50). Bare soil in Kansas can rise as high as 69°C (93), sandy soil in New Hampshire 67°C (126, 224) and sandy soil in Egypt 69°C (137).

During the day the surface of the soil absorbs solar radiation and heat moves downwards through the soil, whereas during the night the soil surface loses heat by radiation and heat moves upwards through the soil. Thus, at a given depth there are, twice in each 24 hours, periods when the direction of heat movement is changing and there is consequently for a while no movement of heat (138). During the day temperature decreases with depth and when the surface temperature is high there may be a drop of some 10°C in the first few centimetres (50). During the night, temperature usually increases from the surface downwards. The greatest daytime gradient is larger than the greatest night-time gradient, and the size of the gradient varies with the time of day or night (23).

Although quantitative differences in temperature phenomena exist between different soil types and climates there is some measure of general similarity. The average day temperature in the soil down to approximately 12 cm is higher than the average night temperature but, owing to the thermal capacity of the soil, between approximately 12 cm and 50 cm the night temperature is higher than the day temperature; below about 60 cm there is no regular diurnal variation (205). The time of the year also exerts an effect, and in summer the average temperature over 24 hours falls with depth, whereas in winter it rises (176). For example, the average temperature at 7 cm below the surface is higher in summer and lower in winter than at a depth of 60 cm (141). Twice a year, in spring and autumn, the soil reaches a uniform temperature throughout, comparable to the spring and autumn overturn in lakes. In North Dakota the spring inversion of soil temperature occurs in early May and the autumn inversion in early October (179). At a depth of approximately 14 m the temperature is constant throughout the year (209, 194). Even at a depth of 1.5-3 m temperature is nearly constant (56). Above this depth the temperature is continually changing.

Near the surface of the soil the daily fluctuation can be very great. For example in Death Valley, California, the average daily range of surface temperature is more than 40°C; in Arizona, on a day when 71°C was recorded 0.4 cm below the surface, the lowest temperature was 17°C (202); and in Egypt temperatures ranging from 66 to 22°C were recorded 1 cm below the surface (242). As might be expected the daily rise and fall in the temperature of the soil surface decreases in amplitude from summer to winter (23, 109), and the daily fluctuation also decreases with depth (56, 204, 210, 211, 217). In India a daily change in temperature occurs deeper than 30 cm; at this depth the daily change averages 1°C (123). In South Africa the daily fluctuation at 60 cm is less than 0.5°C (141), and in England diurnal fluctuations occur down to approximately 60 cm (194).

Because of the thermal capacity of soil, changes in temperature at the soil surface take some hours to affect the temperature at a depth of 15 cm (110). In Nebraska, regardless of the time of year, the highest daily temperature at 7-15 cm below the surface occurs in the evening, whereas at 30 cm it occurs the following morning (217). In summer in Indiana the highest daily temperature 22 cm deep occurs about 10 p.m. and the lowest about 10 a.m. (169). In New York in May the highest temperature 30 cm deep occurs between 8 p.m. and midnight and the lowest between 8 a.m. and noon (134). The lag behind the change in surface temperature is approximately two hours at a depth of 8 cm and eight hours at 45 cm (123). At 90 cm only the annual fluctuation in temperature is measurable (109).

Because of this lag the greater the soil depth the later in the year do the lowest and the highest annual temperatures occur (184). In England at a depth of 1.25 m the highest temperature is reached in August, but at approximately 8 m deep it does not occur until mid winter. Daily fluctuations occur down to approximately 60 cm, between 60 cm and 14 m there are only seasonal fluctuations, and below 14 m there are no fluctuations (194).

Rain markedly affects soil temperature; while rain is falling there is virtually no temperature gradient in the top 30 cm of the soil (50). Rainfall is the only cause of large, nearly simultaneous temperature changes down to a depth of 60 cm (189).

Changes in air temperature are quickly reflected in the soil at a depth of 0.5 cm (176), but the lag in soil temperature behind air temperature increases with depth (176, 204, 210, 211). For example, in Kansas in mid summer the highest temperature at a depth of 30 cm can occur four hours after the highest air temperature (93), and in England in summer the highest temperature 15 cm deep can occur three hours after the highest air temperature (110).

Changes in the temperature of the soil surface can be greater, or as great as, changes in air temperature, but below the surface of the soil the changes are smaller. In Kansas the temperature of the air and the soil surface can fall 22°C in two days, but during this period the temperature at a depth of 30 cm falls only 3°C (93). Below 90 cm, however, soil temperature does not reflect daily changes in air temperature (56). In South Africa the annual range in air temperature can be 40°C, whereas the annual range in soil temperature at 7 cm is 33°C and at 60 cm only 14°C (141).

This survey of naturally occurring soil temperatures shows that temperature is rarely, if ever, constant and normally varies with both time and depth. As has been said, most of the experimental data on the effects of root temperature presented so far have been obtained with constant root temperatures. Therefore, before extrapolating the results obtained experimentally under constant conditions to the likely response of plants growing under fluctuating natural conditions, attention should be given to the effects of changes in root temperature.

3.2. Changes in root temperature

Brouwer (38) grew beans *(P. vulgaris)* at root temperatures of 10 and 20°C for the first ten days after germination. The seedlings were then grown for the next ten days at root temperatures of 10, 15, 20, 25, 30 or 35°C. When the initial temperature was 10°C the rates of fresh weight gain after the change in root temperature were higher at all the higher root temperatures than they were for the plants which remained at a constant 10°C. This might be expected since 10°C is a low root temperature for the growth of beans. When the initial temperature was 20°C the growth rates after the change were similar at 25 and 30°C to those of the plants which remained at a constant 20°C, while those transferred to 10, 15 and 35°C had lower growth rates. This result is also not surprising, because temperatures between 20 and 30°C are close to the optimum whereas 10-15°C is sub-optimal and 35°C is supra-optimal. These responses are clearly predictable from data previously presented in this review, which had been obtained in constant root temperatures.

Tagawa (218), however, also using bean *(P. vulgaris)* seedlings, obtained evidence for an influence of a change in root temperature on stomatal aperture. The seedlings were grown in a constant air temperature, atmospheric humidity and light intensity at a root temperature of 20°C. The root temperature was then either maintained at 20 or changed to 5, 10, 15, 25 or 30°C. The effect of the sudden change in temperature was similar regardless of temperature. The stomata showed a slight temporary closure and then opened wide, and by the second day after the change in root temperature the stomatal apertures at all root temperatures were still greater than the original aperture at 20°C.

A rather similar response has been observed by Grossenbacher (86) in the exudation of sunflowers decapitated just below the cotyledons. The rate of exudation was greater at a root temperature of 30° than at 15°C, but it was highest of all when the temperature was changed from 15 to 30°C. This stimulation continued for two days after the change, but by the third day the rate of exudation had declined to that at 30°C. A similar observation for the absorption of nitrate nitrogen by lemon seedlings was recorded by Vinokur (227), who found that at a constant root temperature of 33°C absorption was greater than at 13° but greatest of all when the temperature was changed from 13 to 33°C. Similarly, Zhurbitzky and Shtrausberg (245) found that oats grown at a root temperature of 10° and then subjected to 18°C absorbed about twice as much phosphorus as oats grown continuously at 18°C. They stated that "the change of temperature in itself produces a strong effect on the absorption of mineral nutrients".

Takeshima (220) observed that the uptake of nitrogen, phosphorus and potassium by rice plants was greater when there was a daily alternation of root temperature than when the temperature was constant. This effect was parallelled by the influence of a daily alternation of root temperature on the yield of grain. Raney, Hagan and Finfrock (186) grew rice plants at constant root temperatures of 18, 21, 27 and 32°C and with a daily alternation of 21 and 27°C. Grain yield increased with constant root temperature, but the yield was highest of all when there was a daily alternation. A similar effect on plant height was observed by Herath and Ormrod (92), who grew rice plants at constant root temperatures of 24 and 32°C and with a daily alternation of 24 and 32°C. Plant height increased with constant root temperature but was greatest when there was a daily alternation.

Gellerman, Litvinenko and Knyazev (81) have also obtained data which might be interpreted as providing evidence for a beneficial effect on tomato growth of a change in root temperature, although other interpretations are possible. They grew tomato plants at a constant air and root temperature of 25°C and raised the root temperature daily to 38°C for 1, 3 or 6 hours either in the morning or evening. Rather surprisingly, raising the temperature to this supra-optimal level increased growth. The greatest increase was obtained when the high temperature was maintained for the shortest time (one hour), and raising the temperature in the morning increased growth more than when it was raised in the evening.

That there is an effect on transpiration of a change in root temperature, which may be attributed to the change itself, is apparent in the work of Nelson (154), who grew cotton seedlings at a root temperature of 28°C, dropped the temperature to 12°C and recorded transpiration over the subsequent four days. The rate of transpiration did not merely change from that at 28°C to another rate at 12°C. It fell and then gradually rose during the four days of observation.

These examples of the influence of a change in root temperature are sufficiently numerous to suggest that it would be wise, when it is intended to extend relationships obtained in controlled environments to the growth of plants in the field, to determine the influence of a change in root temperature on any response curves obtained in constant environmental conditions.

PART 4

PRACTICAL CONTROL OF ROOT TEMPERATURE

One of the reasons for there not having been a greater interest in root temperature among applied biologists is the difficulty of controlling root temperature in crop production. The greatest control is found in the heated glasshouse industry, where techniques such as bench warming (i.e. the growing of plants in pots on benches warmed by electric heating cables) are employed. However, even in heated glasshouses where crops are grown in the border soil, it is the air temperature that is controlled and the root temperature merely follows the air temperature, there being no independent control of root temperature.

Nevertheless, even in the production of field crops it is possible to influence root temperature to some extent. The choice of a site is an obvious factor, particularly the influence on root temperature of altitude and the aspect of the site. The degree of shade that can be obtained from cover plants, the use of soil mulches and the effects of irrigation and soil moisture are all factors that can be employed. Their influence is considered in detail in the following five sections.

4.1. Aspect

As might well be expected, the aspect of a site affects the root temperature of plants grown on it. Thus Pool (177) reported that the temperatures at the soil surface and at depths of 15 and 45 cm were greater on the southern slopes of sandhills in Nebraska than on the northern slopes. Soil temperatures along a north-south transect of a valley running from east to west in Colorado were studied by Bates (15). He found that the highest soil temperature of the year, both at the surface and at a depth of 30 cm, occurred about halfway up the south-facing slope.

In addition, the type of plant cover can modify the influence of aspect on soil temperature. The soil 4 cm below the surface under grassland on the southern slope of a mountain range in New Mexico in summer was, on average, 4°C warmer than it was on the northern slope. Under spruce-fir cover, however, there was little difference between the north and south slopes (80).

When plants are grown in pots the further influence of aspect is superimposed on that of the site. The effect differs according to whether the rows of pots are close together or widely spaced so that they do not shade each other. Pfeiffer and Pettibone (175) observed that the greatest daily temperature fluctuation in pots, standing close together in a block, occurred in the southernmost row. Self and Ward (195) found that when pots were widely spaced the highest temperatures occurred on the south-western sides of the pots, and that loquat plants developed one-sided root systems with fewer roots in the south-western part of the soil.

4.2. Containers

The type of container in which the plant is grown can also influence root temperature. For example, the temperature of the rooting medium in porous containers such as clay pots, is lower than that in non-porous containers such as plastic pots. This is because of the cooling effect of the evaporation of water from the walls of the containers.

Several workers have examined this effect. Lake (122) found a reduction of less than 1°C, Hubbell (96) reported a reduction of about 1°C, Bunt and Kulwiec (42) found a reduction ranging from 1°C in winter to 4°C in summer, and Jones (105) observed a reduction as high as 11°C. Jones (104) also recorded water loss and found the loss from porous containers to be approximately 2.6 times as great as from non-porous containers. Bunt and Kulwiec (42) showed that when evaporative cooling was prevented by coating clay pots with bitumen the temperature reduction was less.

4.3. Cover plants

The use of cover plants in crop production is a well established practice; for example, orchards are grassed down and shade trees are used in tea plantations. There is, consequently, considerable information available on the influence of cover plants on soil temperature.

Considering grass cover first, it is known that in summer in Indiana the highest daily temperature at a depth of 10 cm is greater in bare cultivated soil than under grass but, because of the more rapid cooling of the bare soil at night, there is little difference between the lowest daily temperatures at that depth. At 50 cm depth, however, night cooling has little effect during short summer nights, and consequently the temperature of the bare soil at this lower depth is greater than under grass throughout the 24 hours (32). The more rapid cooling of bare soil at night was also apparent in Japan where, 5 cm below the soil surface in winter, the lowest temperature could be 2°C lower under bare soil than under grass, whereas in summer the highest temperature could be 8°C warmer under the bare soil (99). Thus a grass cover reduces the daily fluctuation in soil temperature (126, 168). During a period of hard freezing the average temperature at a depth of 15 cm can be 1°C higher under grass than under bare cultivation. Frost can penetrate to a depth of 20 cm in cultivated land, when penetration is only 10 cm under adjacent grass (57). The total frost-free period under grass in North Dakota can be a month longer than under bare soil (179). Cutting the grass affects its influence. For example, cutting a sward of *Poa pratensis* lessens its ability to reduce heat absorption by day and heat loss by night, relative to bare soil (40). Prairie grassland in Nebraska that had been undisturbed by mowing, grazing or burning for 15 years, had approximately 15 cm of natural mulch. The temperature just below the soil surface in such a situation could be 14°C below that in bare soil (233). It has also been shown that the temperature just below the surface of grassland with the surface litter burnt off can be higher than in unburned grassland (223).

Considering tree cover next, the soil surface under mature forest cover can be 25°C cooler in summer than the soil surface of sand dunes at a similar altitude. Even at a depth of 45 cm the forest soil could be cooler than the sand dunes. In young plantations, however, the smaller trees do not prevent the sun from shining for a long time on areas between the trees, although they do provide some shelter from wind. In such areas the soil surface temperature can be higher than on the sand dunes (51). Shading dry soil reduces its temperature (61), and shading the sand in the Egyptian desert cooled it to a depth of at least 18 cm (242). The surface of the soil under a forest canopy in South Africa was cooler than on sites freely exposed to sunshine (176). Felling increased the temperature of the layer of forest litter. In a hardwood forest in New Hampshire the highest temperature in the litter in a felled area was 77°C, but only 22°C in an unfelled area (88).

Forest cover reduces the direct sunshine which reaches the ground, and by night it reduces the loss by radiation of heat from the soil surface (200); under forest cover, therefore, the soil temperature fluctuates less than in the open (126), owing both to the screening effect of the canopy and to the insulating effect of the layer of forest litter (234). Even sparse shade from shrubs in New Mexico in summer could lower the temperature of the soil 13 cm below the surface from 57°C to 21°C (214), and the soil 30 and 120 cm below the surface was cooler in summer in a pine plantation in Nebraska than in adjacent open ground. In the winter, however, the soil could be warmer in the pine plantation than in the open ground (157). Surface temperatures under forest canopy in South Africa were lower than on similar exposed sites throughout the year, but the difference was less in winter than in summer (176). In general, the average temperature at similar depths under forest cover is lower than in the open during spring, summer and autumn, but in winter there is little difference. Moreover, at similar depths the average annual temperature under forest cover is lower than in the open (126).

Sometimes the differences are small, between the influences of different types of forest cover. In Mississippi the soil temperatures down to 45 cm below the surface could be very similar in pine plantations, oak forest or basswood-maple forest (51). On the other hand, soil temperatures can be lower in hemlock forest than in hardwood forests of oak, beech and sugar maple (148). At similar depths the soil tends to be cooler in summer and autumn under deciduous forest than under evergreen forest, but in spring it tends to be warmer; in winter there is little difference (126).

There have been some comparisons made between the influence of grass and tree cover on soil temperature. Holch (94) observed that in summer, on the southern slope of a hill in Nebraska, the soil down to a depth of 60 cm was cooler in an oak wood than it was under grass. Pearson (173) found that in summer in Arizona the soil at a depth of 60 cm was cooler in pine forest than it was under grass. The daily fluctuation of temperature under forest cover is also less than it is under grassland (126).

As is to be expected, the actual crops themselves also affect soil temperatures. The presence of a cotton crop kept the soil at a depth of 15 cm cooler than where there was no crop (58), and crop shade reduced the surface temperature in India by 27°C, and at a depth of 20 cm by 2°C (185). Cover crops of lucerne, clover and cowpeas reduced soil temperature (210), but in winter a cover crop could raise the temperature at a depth of at least 15 cm. A maize crop delayed the seasonal penetration of heat into the soil, and the highest temperature of the year at a depth of 180 cm did not occur until nearly two months after its occurrence in bare soil. The seasonal loss of heat was also affected, and the lowest temperature occurred nearly two months later than in bare soil (179).

4.4. Mulching

When the surface soil is cultivated and a loose mulch of dry soil is formed, the heat conductivity of the soil is reduced. Under these conditions a greater proportion of the surface heat is lost by radiation (22) but, during the day, the actual surface temperature of such loose soil can be higher than that of bare consolidated soil (23). Despite this, because of the reduction in conductivity, a loose soil mulch can reduce the temperature during hot dry weather down to a depth of at least 15 cm (141).

Covering the soil with a straw mulch reduces the daily fluctuation in temperature (52, 59, 235). In Tanganyika mulching with banana leaves has a similar effect (82), as does mulching with sugar cane trash in Hawaii (139) and mulching with Guinea grass *(Panicum maximum)* in Brazil (144). Such mulches reduce the daily fluctuation in temperature because they reduce the gain of heat by day and the loss of heat by night (235), but although there is a reduction in the highest temperature, the effect on the lowest temperature is less (139, 113). The soil at a depth of 10 cm under sugar cane trash in Queensland is cooler than under bare land from just before noon to just before midnight, and tends to be warmer from just after midnight to just before noon (133). The soil was cooler under a straw mulch in Ohio (150), South Africa (153), Arizona (210) and Nebraska (132); it was cooler under a mulch of sugar cane trash in Puerto Rico (237) and under a mulch of banana leaves in Tanganyika (82). In Brazil a mulch of Guinea grass can reduce the temperature at a depth of 5 cm by 20°C (144). Straw mulch can, however, in some instances reduce daytime temperatures for only a few hours during the warmest part of sunny days and have little effect on cloudy days. Furthermore, owing to shading, a growing crop will progressively reduce the effect of a mulch.

The effect of a mulch decreases as it decays. A fresh straw mulch reflects more radiation than straw that has darkened in colour with exposure, but even dark straw is more reflective than bare soil (131). A straw mulch tends to cool the soil in summer and keep it warm in winter (59, 169). During winter in Kansas, when bare soil is frozen to a depth of 35 cm, the depth to which frost penetrates can be halved under a straw mulch (12); when the surface temperature of bare soil is 22°C a mulch of straw can raise the soil temperature at a depth of 20 cm by 10°C (100).

Even a layer of snow can act as a mulch and affect soil temperature. Bare ground covered with snow can freeze in Kansas to a depth of 35 cm; if the snow is removed the depth of frost penetration is nearly doubled (12). In areas where wind tends to blow the snow off exposed orchard land, a winter cover crop will raise the soil temperature by retaining the drifting snow (70).

Paper impregnated with bitumen has been used as a mulch. Covering the soil with black paper raises the soil temperature (206, 215), the effect being greatest during a spell of sunny weather (72), but after 2-3 days without sunshine the soil under black paper is no warmer than bare soil (140). Because the effect of black paper is greater when the sun is shining, its influence on temperature is greater in summer than in winter. Moreover, as a crop grows, the effect of the black paper mulch becomes less because of the increasing shade from the crop (215). Covering the soil in California with black paper made the soil warmer both by day and by night (203), and particularly by night (221). Covering the soil with paper of a lighter colour than the soil cooled the soil (206) both by day and by night (203).

Plastic film has more recently been used for soil mulching. Smith (208), Chipman (53), Baskett (14), Thorskrud (222) and Irizarry, Azzan and Woodbury (98) reported that soil temperatures were raised under a black plastic mulch. This effect of black plastic increased with the approach of summer (31), was greater in sunny than in cloudy weather (222) and ceased when crop growth shaded the plastic (66). This change in soil temperature can directly affect growth and yield. For example, Ekern (69) reported that mulching with black plastic in Hawaii had little effect on soil moisture content but raised the soil temperature and, therefore, increased pineapple growth. On the other hand, Austin (9) has stated that a black plastic mulch on outdoor tomatoes in England "conserved soil moisture but had only a slight effect on mean soil temperature. The yield increases were thus most probably brought about by soil moisture rather than by soil temperature effects." Moore (149), who mulched grape vines with black polythene, also concluded that the beneficial effect on growth appeared to be due to moisture conservation because the soil temperature was not increased sufficiently to affect plant growth. Winter survival of maple seedlings was increased by mulching with black plastic because soil heaving by frost was reduced (146).

Voth and Bringhurst (229), Mareček and Trefná (142) and Brock (33) reported that soil temperatures were increased under a clear plastic mulch. The latter also observed that the fluctuation in soil temperature was reduced. Voth and Bringhurst (228) and Peterson, Robbins and Weigle (174) found that daytime soil temperatures were raised and Cannell, Voth and Bringhurst reported that both day and night temperatures were raised (46).

Goodall and Cahoon (83) and Mostert (151) reported that both clear and black plastic film raised the soil temperature. Black, grey, green and clear film all raised winter soil temperatures in Italy and there was no difference between them (30). On the other hand, Downes (66) reported that the soil was warmer under clear than under grey film, and warmer under grey than under black. In India (170), the U.S.A. (114) and France (125) soil temperatures were raised by both black and clear film, but more by clear. Takatori, Lippert and Whiting (219) observed that increases in day temperature were less under black film than under clear, but that black conserved more soil heat by night. Locascio and Smart (129) found that maximum and minimum soil temperatures were higher under clear film than under grey or black. On the other hand, Shadbolt, McCoy and Whiting (196) reported that maximum soil temperatures were higher under clear than under black film but that minimum temperatures were similar. Black film can reduce the maximum temperature in spring and increase it in summer, whereas clear film increased it in both spring and summer (89). In a comparison of clear, black and white plastic mulch and a straw mulch soil temperature was raised most by clear plastic and less by black, was unaffected by white and was reduced by straw (192). Brand and Hawthorne (29) also noted that soil temperature was higher under black plastic than under straw. In a comparison of PVC and polythene film mulches both raised the soil temperature but PVC had a slightly greater effect (41). An ingenious form of mulching was tested by Bowers (24) in Idaho. He compared clear plastic and clear plastic film filled with water. When the air temperature was 24°F the temperatures just below the soil surface were 32°F without plastic mulch, 37°F under clear plastic and 41°F under water-filled plastic.

The possibility that the influence of plastic mulching on crop growth is not always the result of the direct effect of soil temperature, but may also be related to the influence of the plastic film on soil moisture, has already been mentioned. A further complexity is apparent from the work of Miller and Waggoner (147) who mulched young apple trees with clear and black plastic film. The clear plastic raised the soil temperature more than the black, and the soil fungus *Rhizoctonia solani* increased under the black but not under the clear film. They also observed that the nematode *Pratylenchus penetrans* was greatly reduced under black film. They concluded that the black film promoted the biological control of the nematode by soil fungi.

4.5. Soil moisture

Moist soil is usually cooler than dry soil. One reason for this is the high specific heat of water (207). Dry soil has a specific heat that is approximately one-fifth that of water, hence a given amount of heat will raise the temperature of moist soil less than that of dry soil (109), and, consequently, sunshine raises the temperature of moist soil less than that of dry soil (110).

Moreover, the evaporation of moisture from the surface of the soil also cools the soil. This is shown when a thin layer of sand is spread over a dark soil with a high moisture content, such as a peat soil. Although the sand is lighter in colour and would, consequently, absorb less heat than the dark surface, yet the soil under the sand becomes warmer because of the reduction in surface evaporation (23). Thus an irrigation programme that maintains a continually moist soil surface is most effective in cooling the soil (236). In Arizona soil temperature was reduced down to a depth of 90 cm following irrigation, but the effect disappeared within two weeks. However, when there were 7-10 days between irrigations the reduction in soil temperature was maintained (210). Irrigation reduced soil temperature regardless of whether the irrigation water was warmer or cooler than the soil. The greatest reduction occurred 36 hours after irrigation; thereafter the temperature slowly rose until near the end of the second week when there was no further influence on soil temperature (211). For a few days after irrigation soil cropped with cotton in India was cooled and then the temperature rose (58). Watering a sward of *Poa pratensis* reduced both the highest and lowest temperatures (40). Moistening a grey exposed soil in Idaho lowered its surface temperature (61). Moistening dark soil in India with the equivalent of 12 mm of rain reduced its temperature down to a depth of 20 cm, but the effect disappeared after three days (185).

The influence of rain and irrigation water upon short-term changes in soil temperature largely depends on the temperature of the soil just before the application of the water. If the upper layers of the soil are warmer than the lower layers then the upper layers are cooled and the temperature of the lower layer rises. If the upper layers are cooler, then both the upper and the lower layers are cooled by the application of water. The falling of 38 mm of rain, or the application of a similar amount of irrigation water, affects temperature to a depth of at least 50 cm. The least effect occurs if the water is applied when the temperatures of the upper and lower layers of soil are similar. This occurs twice daily, early and late in the day. If the water is applied when the surface temperature is high there is a sudden rise in temperature lower down in the soil, but despite this short-term rise both the highest and lowest temperatures are reduced down to the depth of penetration of the water, because of the increase in specific heat resulting from the addition of the water (136). Applying cold irrigation water to warm soil lowers the temperature at a depth of 10 cm to nearly the same as the temperature of the irrigation water, but within 24 hours the temperature rises considerably (240). In South Africa rain in summer lowered the temperature for less than one week. This effect of rain decreased with depth, and during the cooler winter months had less effect on soil temperature (141).

BIBLIOGRAPHY

1 Abd El Rahman, A. A., Kuiper, P. J. C. & Bierhuizen, J. F. (1959). *Meded. LandbHogesch. Wageningen,* **59**(15): 1-12. Preliminary observations on the effect of soil temperature on transpiration and growth of young tomato plants under controlled conditions.

2 Adams, W. R. (1934). *Bull. Vt agric. Exp. Stn,* No. 379, 18 pp. Studies in tolerance of New England forest trees. XI. The influence of soil temperature on the germination and development of white pine seedlings.

3 Allen, R. C. (1934). *Proc. Am. Soc. hort. Sci.,* **32:** 635-637. The effects of soil temperature on the growth and flowering of certain greenhouse crops.

4 Anderson, W. B. & Kemper, W. D. (1964). *Agron. J.,* **56:** 453-456. Corn growth as affected by aggregate stability, soil temperature and soil moisture.

5 Andreenko, S. S. & Kerechki, B. (1966). *Dokl. Akad. Nauk. S.S.S.R.,* **168:** 101-103. [On the effect of a reduced temperature in the root zone on certain physiological processes in maize plants.]

6 Army, T. J. & Miller, E. V. (1957). *Proc. Soil. Sci. Soc. Am.,* **21:** 176-182. Relationships of the cation suite and yield of turnip greens for selected extremes of soils, fertilizers, and environment.

7 Arndt, C. H. (1932). *A.R.S. Carol. agric. Exp. Stn,* **45:** 46-49. A study of some of the factors which may influcnce cotton seed germination and seedling growth.

8 Ashby, W. C. (1960). *Bot. Gaz.,* **121:** 228-233. Seedling growth and water uptake by *Tilia americana* at several root temperatures.

9 Austin, R. B. (1964). *Exp. Hort.,* **11:** 17-22. Plastic mulches for outdoor tomato crops and a trial of varieties.

10 Babalola, O., Boersma, L. & Youngberg, C. T. (1968). *Pl. Physiol.,* **43:** 515-521. Photosynthesis and transpiration of Souterey pine seedlings as a function of soil water suction and soil temperature.

11 Bailey, J. S. & Jones, L. H. (1941). *Proc. Am. Soc. hort. Sci.,* **38:** 462-464. The effect of soil temperature on the growth of cultivated blueberry bushes.

12 Barnett, R. J. (1944). *Proc. Am. Soc. hort. Sci.,* **44:** 57-65. Effect of ground cover on the freezing and thawing of orchard soils.

13 Barney, C. W. (1951). *Pl. Physiol.,* **26:** 146-163. Effects of soil temperature and light intensity on root growth of loblolly pine seedlings.

14 Baskett, W. J. (1960). *J. Agric. S. Aust.,* **64:** 149. Polythene plastic sheeting for mulching fruit trees and vegetables.

15 Bates, C. G. (1923). *Ecology,* **4:** 54-62. The transect of a mountain valley.

16 Bates, C. G. (1926). *Ecology,* **7:** 469-480. Some relations of plant ecology to silvicultural practice.

17 Beauchamp, E. G. & Lathwell, D. J. (1966). *Can. J. Pl. Sci.,* **46:** 593-601. Effect of root zone temperatures on corn leaf morphology.

18 Benedict, H. M. (1950). *Pl. Physiol.,* **25:** 377-388. The effect of soil temperature on guayule plants.

19 Bewley, W. F. (1934). *J. Minist. Agric. Fish.,* **40:** 1047-1056. Raising the soil temperatures in glasshouses.

20 Bialoglowski, J. (1936). *Proc. Am. Soc. hort. Sci.,* **34:** 96-102. Effect of extent and temperature of roots on transpiration of rooted lemon cuttings.

21 Böhning, R. H. & Lusanandana, B. (1952). *Pl. Physiol.,* **27:** 475-488. A comparative study of gradual and abrupt changes in root temperature on water absorption.

22 Bouyoucos, G. J. (1915). *Tech. Bull. Mich. agric. Exp. Stn,* No. 22, 63 pp. Effect of temperature on the movement of water vapour and capillary moisture in soil.

23 Bouyoucos, G. J. (1916). *A. R. Mich. agric. Exp. Stn,* **29:** 688-816. Soil temperature.

24 Bowers, S. A. (1967). *Idaho agric. Sci.,* **52**(3): 6. Water filled bags make good mulch.

25 Boxall, M. I. (1962). *Rep. elect. Res. Ass.,* No. W/T43, 14 pp. Some effects of soil temperature on plant growth.

26 Boxall, M. I. (1962). *Shinfield Prog.,* No. 2, pp. 19-23. Soil temperature and plant growth.

27 Boxall, M. I. (1963). *Shinfield Prog.,* No. 3, p. 19. Soil temperature and plant growth.

28 Boxall, M. I. (1964). *Shinfield Prog.,* No. 6, p. 35. The effect of soil temperature on the growth of the chrysanthemum.

29 Brand, H. J. & Hawthorne, P. L. (1956). *Bull. La St. Univ.,* No. 591, 40 pp. Cold protection for Louisiana strawberries.

30 Branzanti, E. C. & Celotti, C. (1967). *Inf. agrar., Verona,* **23:** 2085-2086. Influenza del foglio di plastica di diverso colore sul microclima di parcelle pacciamate.

31 Braud, H. J., Jr. & Chesness, J. L. (1969-70). *La Agric.,* **13**(2): 12-14. Studies show temperature effects of mulch.

32 Brawand, H. & Kohnke, H. (1952). *Proc. Soil. Sci. Soc. Am.,* **16:** 195-198. Microclimate and vapor exchange at the soil surface.

33 Brock, J. R. (1965). *Parks,* **30:** 512, 514. Grass under plastic.

34 Brouwer, R. (1959). *Jaarb. Inst. biol. scheik. Onderz. LandbGewass. 1959,* pp. 27-36. De invloed van de worteltemperatuur op de groei van erwten.

35 Brouwer, R. (1961). *Jaarb. Inst. biol. scheik. Onderz. LandbGewass. 1961,* pp. 11-24. Water transport through the plant.

36 Brouwer, R. (1962). *Jaarb. Inst. biol. scheik. Onderz. LandbGewass. 1962,* pp. 11-18. Influence of temperature of the root medium on the growth of seedlings of various crop plants.

37 Brouwer, R. (1963). *Jaarb. Inst. biol. scheik. Onderz. LandbGewass. 1963,* pp. 11-30. Some physiological aspects of the influence of growth factors in the root medium on growth and dry matter production.

38 Brouwer, R. (1964). *Jaarb. Inst. biol. scheik. Onderz. LandbGewass. 1964*, pp. 11-22. Responses of bean plants to root temperatures. 1. Root temperatures and growth in the vegetative stage.

39 Brouwer, R. & Vliet, G. van (1960). *Jaarb. Inst. biol. scheik. Onderz. LandbGewass. 1960*, pp. 23-36. The influence of root temperature on growth and uptake of peas.

40 Brown, E. M. (1943). *Res. Bull. Mo. agric. Exp. Stn*, No. 360, 56 pp. Seasonal variations in the growth and chemical composition of Kentucky bluegrass.

41 Brun, R. (1968). *Pépeniéristes, Horticulteurs, Maraîchers*, No. 86, pp. 4851-4859. Bilan de six années d'étude et d'emploi du paillage plastique en France.

42 Bunt, A. C. & Kulwiec, Z. J. (1970). *Plant & Soil.*, **32:** 65-80. The effect of container porosity on root environment and plant growth. 1. Temperature.

43 Burkholder, W. H. (1920). *Ecology*, **1:** 113-123. The effect of two soil temperatures on the yield and water relations of healthy and diseased bean plants.

44 Calvert, A. (1956). *J. hort. Sci.*, **31:** 69-75. The influence of soil and air temperatures on the cropping of glasshouse tomatoes.

45 Camp, A. F. & Walker, M. N. (1927). *Bull. Fla agric. Exp. Stn*, No. 189, 32 pp. Soil temperature studies with cotton.

46 Cannell, G. H., Voth, V. & Bringhurst, R. S. (1961). *Proc. Amer. Soc. hort. Sci.*, **78:** 281-291. The influence of irrigation levels and application methods, polyethylene mulch and nitrogen fertilization on strawberry production in Southern California.

47 Cannon, W. A. (1917). *Yb. Carnegie Instn Wash.*, **15:** 75-76. Rate of root-growth of *Covillea tridentata* in relation to the temperature of the soil.

48 Cannon, W. A. (1918). *Pl. World*, **21:** 64-67. The evaluation of the soil temperature factor in root growth.

49 Carlson, R. F. (1965). *Proc. Am. Soc. hort. Sci.*, **86:** 41-45. Responses of Malling Merton clones and Delicious seedlings to different root temperatures.

50 Chapman, R. N., Mickel, C. E., Parker, J. R., Miller, G. E. & Kelly, E. G. (1926). *Ecology*, **7:** 416-426. Studies in the ecology of sand dune insects.

51 Chapman, R. N., Wall, R., Garlough, L. & Schmidt, C. T. (1931). *Ecology*, **12:** 305-322. A comparison of temperatures in widely different environments of the same climatic area.

52 Chiba, T., Sekiya, K., Oaba, K. & Suzuki, K. (1967). *Bull. hort. Res. Stn, A (Hiratsuka)*, **6:** 9-27. [Studies on soil management. 8. The effects of some organic mulches on the behaviour of young peach trees.]

53 Chipman, E. W. (1961). *Canad. J. Plant Sci.*, **41:** 10-15. Studies of tomato response to mulching on ridged and flat rows.

54 Clements, F. E. & Martin, E. V. (1934). *Pl. Physiol.*, **9:** 619-630. Effect of soil temperature on transpiration in *Helianthum annuus*.

55 Cox, L. M. & Boersma, L. (1967). *Pl. Physiol.*, **42:** 550-556. Transpiration as a function of soil temperature and soil water stress.

56 Crabb, G. A., Jr. & Smith, J. L. (1953). *Bull. Highw. Res. Bd*, No. 71, pp. 32-80. Soil temperature comparisons under varying covers.

57 Craig, J. (1900). *Bull. Ia agric. Exp. Stn*, **44:** 179-213. Observations and suggestions on the root-killing of fruit trees.

58 Dabral, B. M. & Chiney, S. S. (1938). *Indian J. agric. Sci.*, **8:** 161-184. Micro-climatology of an irrigated cotton field in Sind.

59 Darrow, G. M. & Magness, J. R. (1938). *Proc. Am. Soc. hort. Sci.*, **36:** 481-484. Investigations on mulching red raspberries.

60 Darrow, R. A. (1939). *Bot. Gaz.*, **101:** 109-127. Effects of soil temperature, pH and nitrogen nutrition on the development of *Poa pratensis*.

61 Daubenmire, R. F. (1943). *J. For.*, **41:** 601-603. Temperature gradients near the soil surface with reference to techniques of measurement in forest ecology.

62 Davidson, O. W. (1941). *Proc. Am. Soc. hort. Sci.*, **39:** 387-390. Effects of temperature on growth and flower production of gardenias.

63 Davidson, R. L. (1969). *Ann. Bot.*, **33:** 561-569. Effects of root/leaf temperature differentials on root/shoot ratios in some pasture grasses and clover.

64 Dibbern, J. C. (1947). *Bot. Gaz.*, **109:** 44-58. Vegetative responses of *Bromus inermis* to certain variations in environment.

65 Dickson, J. G. (1923). *Proc. Ass. off. Seed Analysts N. Am. 1921*, **14:** 68-72. The influence of soil temperature and moisture on the germination of wheat treated with different seed treatments.

66 Downes, J. D. (1966). *Hort. Rep. Mich. St. Univ.*, No. 30, p. 5. Comparisons of plastics and fumigation for tomatoes and melons.

67 Duncan, H. F. & Cooke, D. A. (1932). *Hawaii. Plrs' Rec.*, **36:** 31-40. A preliminary investigation of the effect of temperature on root absorption of the sugar cane.

68 Ehrler, W. L. (1963). *Agron. J.*, **55:** 363-366. Water absorption of alfalfa as affected by low root temperature and other factors of a controlled environment.

69 Ekern, P. C. (1967). *Proc. Soil Sci. Soc. Amer.*, **31:** 270-275. Soil moisture and soil temperature changes with the use of black vapor-barrier mulch and their influence on pineapple growth in Hawaii.

70 Emerson, R. A. (1906). *Bull. Neb. agric. Exp. Stn*, No. 92, 23 pp. Cover crops for young orchards.

71 Erickson, L. C. & Smith, P. F. (1947). *Tech. Bull. U.S. Dep. Agric.*, No. 924, 58 pp. Studies on handling and transplanting guayule nursery stock.

72 Flint, L. H. (1928). *Tech. Bull. U.S. Dep. Agric.*, No. 75, 20 pp. Crop-plant stimulation with paper mulch.

73 Forster, H. C. (1935). *J. Aust. Inst. agric. Sci.*, **1:** 27-28. Response of English and Australian wheats to length of day and temperature.

74 Franco, C. M. (1958). *Bull. IBEC Res. Inst.*, No. 16, 21 pp. Influence of temperature on growth of coffee plant.

75 Fritschen, L. J. & Shaw, R. H. (1961). *Agron. J.*, **53:** 71-74. Transpiration and evapotranspiration of corn as related to meteorological factors.

76 Fujishige, N. & Sugiyama, T. (1968). *J. Jap. Soc. hort. Sci.*, **37:** 221-226. [Effect of soil temperature on growth of seedlings of a few fruit vegetables. Preliminary report.]

77 Fulton, J. M. (1968). *Can. J. Soil Sci.*, **48:** 1-5. Growth and yield of oats as influenced by soil temperature, ambient temperature and soil moisture supply.

78 Galligar, G. C. (1938). *Pl. Physiol.*, **13:** 835-844. Temperature effects upon the growth of excised root tips.

79 Garwood, E. W. (1968). *J. Br. Grassld Soc.*, **23:** 117-128. Some effects of soil-water conditions and soil temperature on the roots of grasses and clover. 2. Effects of variation in the soil-water content and in soil temperature on root growth.

80 Gary, H. L. (1968). *Res. Note U.S. For. Serv.*, No. RM-118, 11 pp. Soil temperatures under forest and grassland cover types in northern New Mexico.

81 Gellerman, Y. M., Litvinenko, L. A. & Knyazev, A. N. (1963). *Izv. timiryazev. sel.-khoz. Akad.*, **50:** 38-49. [Stimulation of tomato plant growth by the periodical action of sub-lethal temperatures on the roots.]

82 Gilbert, S. M. (1945). *E. Afr. agric. J.*, **11**(2): 75-79. The mulching of *Coffea arabica.*

83 Goodall, G. E. & Cahoon, G. (1963). *West Fruit Gr*, **17**(5): 40. Tests boost plastic mulch for citrus nursery stock.

84 Gray, G. F. (1941). *Proc. Am. Soc. hort. Sci.*, **39:** 269-273. Transpiration in strawberries as affected by root temperature.

85 Grobbelaar, W. P. (1963). *Meded. LandbHogesch. Wageningen*, **63**(5): 1-71. Responses of young maize plants to root temperatures.

86 Grossenbacher, K. A. (1939). *Am. J. Bot.*, **26:** 107-109. Autonomic cycle of rate of exudation of plants.

87 Guinn, G. & Hunter, R. E. (1968). *Crop. Sci.*, **8:** 67-70. Root temperature and carbohydrate status of young cotton plants.

88 Hall, R. C. (1933). *Bull. Sch. For. Conserv. Univ. Mich.*, No. 3, 66 pp. Post-logging decadence in northern hardwoods.

89 Harris, R. E. (1965). *Proc. Amer. Soc. hort. Sci.*, **87:** 288-294. Polyethylene covers and mulches for corn and bean production in northern regions.

90 Heinrichs, D. H. & Nielsen, K. F. (1966). *Can. J. Pl. Sci.*, **46:** 291-298. Growth response of alfalfa varieties of diverse genetic origin to different root zone temperatures.

91 Hellmers, H. (1963). *Bot. Gaz.*, **124:** 172-177. Effects of soil and air temperatures on growth of redwood seedlings.

92 Herath, W. & Ormrod, D. P. (1965). *Agron. J.,* **57:** 373-376. Some effects of water temperature on the growth and development of rice seedlings.

93 Hide, J. C. (1942). *Proc. Soil Sci. Am.,* **7:** 31-35. A graphic presentation of temperatures in the surface foot of soil in comparison with air temperatures.

94 Holch, A. E. (1931). *Ecology,* **12:** 259-298. Development of roots and shoots of certain deciduous tree seedlings in different forest sites.

95 Hori, Y., Arai, K., Hosoya, T. & Oyamada, M. (1968). *Bull. hort. Res. Stn A (Hiratsuka),* **7:** 187-214. [Studies on the effects of root temperature and its combinations with air temperature on the growth and nutrition of vegetable crops. 1. Cucumber, tomato, turnip and snap bean.]

96 Hubbell, D. S. (1930). *Proc. Am. Soc. hort. Sci.,* **27:** 475-477. The influence of various plant containers on the growth and development of *Pelargonium hortorum.*

97 Humphries, E. C. (1963). *Ann. Bot.,* **27:** 175-183. Dependence of net assimilation on root growth of isolated leaves.

98 Irizarry, H., Azzam, H. & Woodbury, R. (1968). *J. Agric. Univ. P.R.,* **52:** 53-63. Evaluation of black polythene plastic mulch for tomato production in Puerto Rico.

99 Itakura, T. & Shimura, I. (1964). *Bull. hort. Res. Stn, A. (Hiratsuka),* **3:** 1-25. [Studies on orchard soil management. 4. On some physical properties of the soil under different soil management systems.]

100 Iverson, V. E. (1939). *Tech. Bull. Minn. agric. Exp. Stn,* No. 135, pp. 19-30. Studies on some factors relating to hardiness in the strawberry. 2. Winter soil temperatures as a factor in the environment of the strawberry and some other herbaceous plants.

101 Jensen, G. (1960). *Physiol. Pl.,* **13:** 822-830. Effects of temperature and shifts in temperature on the respiration of intact root systems.

102 Johnson, J. & Hartman, R. E. (1919). *J. agric. Res.,* **17:** 41-86. Influence of soil environment on the root-rot of tobacco.

103 Jones, F. R. & Tisdale, W. B. (1921). *J. agric. Res.,* **22:** 17-32. Effect of soil temperature upon the development of nodules on the roots of certain legumes.

104 Jones, L. H. (1931). *Bull. Mass. agric. Exp. Stn,* **277:** 148-161. Flower pot composition and its effect on plant growth.

105 Jones, L. H. (1931). *J. agric. Res.,* **42:** 375-378. Effect of the structure and moisture of plant containers on the temperature of their soil contents.

106 Jones, L. H. (1938). *J. agric. Res.,* **57:** 611-621. Relation of soil temperature to chlorosis of gardenia.

107 Jones, L. R., Johnson, J. & Dickson, J. G. (1926). *Res. Bull. agric. Exp. Stn Univ. Wis.,* No. 71, 144 pp. Wisconsin studies upon the relation of soil temperature to plant disease.

108 Jones, L. R., McKinney, H. H. & Fellows, H. (1922). *Res. Bull. agric. Exp. Stn Univ. Wis.,* No. 53, 35 pp. The influence of soil temperature on potato scab.

109 Keen, B. A. (1932). *Quart. J. roy. met. Soc.,* **58:** 229-250. Soil physics in relation to meterology.

110 Keen, B. A. & Russell, E. J. (1921). *J. agric. Sci.,* **11:** 211-239. The factors determining soil temperature.

111 Ketcheson, J. W. (1968). *Proc. Soil Sci. Soc. Am.,* **32:** 531-534. Effect of controlled air and soil temperature and starter fertilizer on growth and nutrient composition of corn *(Zea mays* L.).

112 Ketellapper, H. J. (1960). *Physiol. Pl.,* **13:** 641-647. The effect of soil temperature on the growth of *Phalaris tuberosa.*

113 King, N. J. (1935). *Cane Grs' quart. Bull.,* **3:** 59-63. Frost damage in cane.

114 Knavel, D. E. & Mohr, H. C. (1967). *Proc. Amer. Soc. hort. Sci.,* **91:** 589-597. Distribution of roots of four different vegetables under paper and polyethylene mulches.

115 Knoll, H. A., Brady, N. C. & Lathwell, D. J. (1964). *Agron. J.,* **56:** 145-147. Effect of soil temperature and phosphorus fertilization on the growth and phosphorus content of corn.

116 Knoll, H. A., Lathwell, D. J. & Brady, N. C. (1964). *Proc. Soil Sci. Soc. Am.,* **28:** 400-403. The influence of root zone temperature on the growth and contents of phosphorus and anthocyanin of corn.

117 Korovin, A. I. & Barskaya, T. A. (1963). *Soviet Pl. Physiol.,* **9:** 331-333. The effect of soil temperature on respiration and activity of oxidative enzymes of roots in cold resistant and thermophilic plants.

118 Kozlowski, T. T. (1943). *Pl. Physiol.,* **18:** 252-260. Transpiration rates of some forest tree species during the dormant season.

119 Kramer, P. J. (1940). *Pl. Physiol.,* **15:** 63-79. Root resistance as a cause of decreased water absorption by plants at low temperatures.

120 Kramer, P. J. (1942). *Am. J. Bot.,* **29:** 828-832. Species differences with respect to water absorption at low soil temperatures.

121 Kuiper, P. J. C. (1964). *Meded. LandbHogesch. Wageningen,* **64**(4): 1-11. Water uptake of higher plants as affected by root temperature.

122 Lake, J. V. (1961). *J. agric. Engng Res.,* **6:** 64-71. A comparison between small flower pots made of clay or plastics.

123 Leather, J. W. (1915). *Mem. Dep. Agric. India, chem. Ser.,* **4**(2): 18-84. Soil temperatures.

124 Levesque, M. & Ketcheson, J. W. (1963). *Can. J. Pl. Sci.,* **43:** 355-360. The influence of variety, soil temperature, and phosphorus fertilizer on yield and phosphorus uptake by alfalfa.

125 Leyvraz, H. & Simon, J. L. (1961). *Rev. romande Agric. Vitic.,* **17:** 27-28. La rechauffement du sol d'une pépinière viticole.

126 Li, T. (1926). *Bull. Sch. For. Yale Univ.,* No. 18, 92 pp. Soil temperature as influenced by forest cover.

127 Liebig, G. F. & Chapman, H. D. (1963). *Proc. Am. Soc. hort. Sci.,* **82:** 204-209. The effect of variable root temperatures on the behavior of young Navel orange trees in a greenhouse.

128 Lingle, J. C. & Davis, R. M. (1959). *Proc. Am. Soc. hort. Sci.,* **73:** 312-322. The influence of soil temperature and phosphorus fertilization on the growth and mineral absorption of tomato seedlings.

129 Locascio, S. J. & Smart, G. C. (1969). *Proc. Fla. St. hort. Soc., 1968,* **81:** 147-153. Influence of polyethylene mulch colors and soil fumigants on strawberry production.

130 Locascio, S. J. & Warren, G. F. (1960). *Proc. Am. Soc. hort. Sci.,* **75:** 601-610. Interaction of soil temperature and phosphorus on growth of tomatoes.

131 McCalla, T. M. (1943). *Proc. Soil Sci. Soc. Am.,* **8:** 258-262. Changes in the physical properties of straw during the early stages of decomposition.

132 McCalla, T. M. (1943). *Trans. Kans. Acad. Sci.,* **46:** 52-56. Microbiological studies of the effect of straw used as a mulch.

133 McCalla, T. M. & Duley, F. L. (1946). *J. Am. Soc. Agron.,* **38:** 75-89. Effect of crop residues on soil temperature.

134 MacDougal, D. T. (1902). *J. N.Y. bot. Gdn,* **3:** 125-131. The temperature of the soil.

135 Mack, A. R. (1965). *Can. J. Soil Sci.,* **45:** 337-346. Effect of soil temperature and moisture on yield and nutrient uptake by barley.

136 McKenzie-Taylor, E. (1924). *Bull. Minist. Agric. Egypt tech. scient. Serv.,* No. 53, 18 pp. The effect of irrigation upon soil temperatures.

137 McKenzie-Taylor, E. & Chamley-Burns, A. (1924). *Bull. Minist. Agric. Egypt tech. scient. Serv.,* No. 31, 46 pp. Soil temperatures during the sharaqui period and their agricultural significance.

138 McKenzie-Taylor, E. & Williams, C. B. (1924). *Bull. Minist. Agric. Egypt tech. sci. Serv.,* No. 40, 24 pp. A comparison of sand and soil temperatures in Egypt.

139 Magistad, O. C., Farden, C. A. & Baldwin, W. A. (1935). *J. Am. Soc. Agron.,* **27:** 813-825. Bagasse and paper mulches.

140 Magruder, R. (1930). *Bull. Ohio agric. Exp. Stn,* No. 447, 60 pp. Paper mulch for the vegetable garden: its effect on plant growth and on soil moisture, nitrates and temperature.

141 Malherbe, I. de V. (1938). *Sci. Bull. Dep. Agric. For. S. Afr.,* No. 174, 28 pp. Soil climate with special reference to temperature fluctuations in an orchard soil at Stellenbosch.

142 Mareček, J. & Trefná, E. (1961). *Rostlinná Výroba,* **7:** 663-678. [Mulching vegetables with polythene film.]

143 Martin, G. C. & Wilcox, G. E. (1963). *Proc. Soil Sci. Soc. Am.,* **27:** 565-567. Critical soil temperature for tomato plant growth.

144 Medcalf, J. C. (1956). *Bull. IBEC Res. Inst.,* No. 12, 47 pp. Preliminary study on mulching young coffee in Brazil.

145 Meyer, M. M., Jr. & Tukey, H. B., Jr. (1967). *Proc. Am. Soc. hort. Sci.,* **90:** 440-446. Influence of root temperature and nutrient applications on

root growth and mineral content of *Taxus* and *Forsythia* plants during the dormant season.

146 Miller, P. M. & Waggoner, P. E. (1962). *Amer. Nursym.*, **115**(5): 150. Increasing plant survival with plastic mulches.

147 Miller, P. M. & Waggoner, P. E. (1963). *Plant & Soil,* **18:** 45-52. Interaction of plastic mulch, pesticides and fungi in the control of soil-borne nematodes.

148 Moore, B., Richards, H. M., Gleason, H. A. & Stout, A. B. (1924). *Bull. N.Y. bot. Gdn,* **12:** 325-350. Hemlock and its environment: field records.

149 Moore, R. C. (1963). *Biokemia,* **1:** 21-23. Plastic mulch aids growth of young grape vines and cuttings.

150 Moore, R. P. (1943). *J. Am. Soc. Agron.,* **35:** 370-381. Seedling emergence of small-seeded legumes and grasses.

151 Mostert, M. (1965). *Groent. en Fruit,* **20:** 1631. Grond afdekken bij meloenen.

152 Nagai, T. & Matsushita, E. (1963). *Proc. Crop Sci. Soc. Japan,* **31:** 385-388. Physico-ecological characteristics in roots of rice plants grown under different soil temperature conditions. 1. Their ecological characteristics.

153 Nel, R. I. (1947). *Fmg S. Afr.,* **22:** 1146-1147. Studies on soil moisture and irrigation practices.

154 Nelson, L. E. (1967). *Agron. J.,* **59:** 391-395. Effect of root temperature variation on growth and transpiration of cotton *(Gossypium hirsutum L.)* seedlings.

155 Nelson, S. H. & Tukey, H. B. (1955). *Quart. Bull. Mich. agric. Exp. Stn,* **38:** 46-51. Root temperature affects the performance of East Malling apple rootstocks.

156 Nelson, S. H. & Tukey, H. B. (1956). *J. hort. Sci.,* **31:** 55-63. Effects of controlled root temperatures on the growth of East Malling rootstocks in water culture.

157 Niederhof, C. H. & Stahelin, R. (1942). *J. For.,* **40:** 244-248. Climatic conditions within and adjacent to a forest plantation in the Nebraska sandhills.

158 Nielsen, K. F. & Cunningham, R. K. (1964). *Proc. Soil Sci. Soc. Am.,* **28:** 213-218. The effects of soil temperature and form and level of nitrogen on growth and chemical composition of Italian ryegrass.

159 Nielsen, K. F., Halstead, R. L., MacLean, A. J., Bourget, S. J. & Holmes, R. M. (1961). *Proc. Soil Sci. Soc. Am.,* **25:** 369-372. The influence of soil temperature on the growth and mineral composition of corn, bromegrass and potatoes.

160 Nielsen, K. F., Halstead, R. L., MacLean, A. J., Holmes, R. M. & Bourget, S. J. (1960). *Can. J. Soil Sci.,* **40:** 255-263. The influence of soil temperature on the growth and mineral composition of oats.

161 Nielsen, K. F., Halstead, R. L., MacLean, A. J., Holmes, R. M. & Bourget, S. J. (1961). In *Proc. 8th int. Grassland Congr. 1960,* pp. 287-292.

Effects of soil temperature on the growth and chemical composition of lucerne.

162 Nightingale, G. T. (1935). *Bot. Gaz.,* **96:** 581-637. Effects of temperature on growth, anatomy and metabolism of apple and peach roots.

163 North, C. P. & Wallace, A. (1955). *Calif. Agric.,* **9**(11): 13. Soil temperature and citrus.

164 O'Leary, J. W. (1965). *Bot. Gaz.,* **126:** 108-115. Root-pressure exudation in woody plants.

165 O'Leary, J. W. (1966). *Ann. Bot.,* **30:** 419-423. Temperature effects on root pressure exudation.

166 Ongun, A. R. & Wallace, A. (1958). *Spec. Rep. Dep. hort. Sci. Univ. Calif.,* No. 1, pp. 75-83. Inorganic composition of small Washington Navel orange trees on different rootstocks grown in a glasshouse with different root temperatures.

167 Ongun, A. R. & Wallace, A. (1958). *Spec. Rep. Dep. hort. Sci. Univ. Calif.,* No. 1, pp. 87-103. Transpiration rates of small Washington Navel orange trees grown in a glasshouse with different rootstocks and at different root temperatures.

168 Oskamp, J. (1915). *J. agric. Res.,* **5:** 173-180. Soil temperatures as influenced by cultural methods.

169 Oskamp, J. (1917). *Proc. Am. Soc. hort. Sci.,* **14:** 118-126. The role of soil temperature in tree growth.

170 Panje, R. R., Gill, P. S. & Alam, M. (1964). *Indian Sugarcane J.,* **9:** 6-8. Studies on conservation of moisture for sugarcane crop. 2. Use of polyethylene film as soil cover.

171 Parups, E. V. & Nielsen, K. F. (1960). *Can. J. Pl. Sci.,* **40:** 281-287. The growth of tobacco at certain soil temperature and nutrient levels in the greenhouse.

172 Parups, E. V., Nielsen, K. F. & Bourget, S. J. (1960). *Can. J. Pl. Sci.,* **40:** 516-523. The growth, nicotine and phosphorus content of tobacco grown at different soil temperature, moisture and phosphorus levels.

173 Pearson, G. A. (1913). *Mon. Weath. Rev. U.S. Dep. Agric.,* **41**(10): 1615-1629. A meteorological study of parks and timbered areas in the western yellow-pine forest of Arizona and New Mexico.

174 Peterson, L. E., Robbins, M. L. & Weigle, J. L. (1970). *Ia Fm Sci.,* **25**(3): 34-35. Mulching and transplanting increase early yield of muskmelons.

175 Pfeiffer, C. & Pettibone, A. (1968). *Comb. Proc. int. Pl. Prop. Soc.,* **17:** 78-85. Soil temperature conditions in container-grown plants.

176 Phillips, J. F. V. (1931). *Mem. bot. Surv. S. Afr.,* No. 14, 327 pp. Forest succession and ecology in the Knysna region.

177 Pool, R. J. (1914). *Minn. bot. Stud.,* **4:** 185-312. A study of the vegetation of the sandhills of Nebraska.

178 Post, K. & Mastalerz, J. (1952). *Bull. N.Y. St. Flower Growers,* **80:** 4. Poinsettias and low soil temperature.

179 Potter, L. D. (1956). *Ecology,* **37:** 62-70. Yearly soil temperatures in eastern North Dakota.

180 Proebsting, E. L. (1943). *Proc. Am. Soc. hort. Sci.,* **43:** 1-4. Root distribution of some deciduous trees in a California orchard.

181 Proebsting, E. L. (1957). *Proc. Am. Soc. hort. Sci.,* **69:** 278-281. The effect of soil temperature on the mineral nutrition of the strawberry.

182 Rajan, A. K. (1966). *J. exp. Bot.,* **17:** 1-19. The effect of root temperatures on water and sulphate absorption in intact sunflower plants.

183 Raleigh, G. J. (1941). *Proc. Am. Soc. hort. Sci.,* **38:** 487-488. The effect of culture solution temperature on water intake and wilting of the musk-melon.

184 Rambaut, A. A. (1900). *Phil. Trans. roy. Soc., A,* **195:** 235-258. Underground temperatures at Oxford in the year 1899, as determined by five platinum-resistance thermometers.

185 Ramdas, L. A. & Dravid, R. K. (1936). *Proc. nat. Inst. Sci. India,* **2:** 131-143. Soil temperatures in relation to other factors controlling the dispersal of solar radiation at the earth's surface.

186 Raney, F. C., Hagan, R. M. & Finfrock, D. C. (1957). *Calif. Agric.,* **11**(4): 19-20. Water temperature in irrigation.

187 Richards, B. L. (1921). *J. agric. Res.,* **21:** 459-482. Pathogenicity of *Corticium vagum* on the potato as affected by soil temperature.

188 Riethmann, O. (1933). *Ber. schweiz. bot. Ges.,* **42:** 152-168. Der Einfluss der Bodentemperatur auf das Wachstum und die Reifezeit der Tomaten.

189 Riley, J. A., Jr. (1957). *Mon. Weath. Rev. U.S. Dep. Agric.,* **85:** 393-400. Soil temperatures as related to corn yield in central Iowa.

190 Roberts, A. N. & Kenworthy, A. L. (1956). *Proc. Am. Soc. hort. Sci.,* **68:** 157-168. Growth and composition of the strawberry plant in relation to root temperature and intensity of nutrition.

191 Sachs, J. von (1887). Oxford: Clarendon Press, pp. 264-265, In *Lectures in the physiology of plants.*

192 Schales, F. D. & Sheldrake, R. (1966). *Proc. Amer. Soc. hort. Sci.,* **88:** 425-430. Mulch effects on soil conditions and muskmelon response.

193 Schroeder, R. A. (1939). *Res. Bull. Mo. agric. Exp. Stn.,* No. 309, 27 pp. The effect of root temperature upon the absorption of water by the cucumber.

194 Searle, S. A. (1969). *Gdnrs' Chron.,* **165**(23): 8-9. Environment and plant life. 1. Soil temperature–how to use it.

195 Self, R. L. & Ward, H. S., Jr. (1956). *Pl. Dis. Reptr,* **40:** 957-960. Effects of high soil temperature on root growth of loquat seedlings in nursery containers.

196 Shadbolt, C. A., McCoy, O. D. & Whiting, F. L. (1962). *Hilgardia,* **32:** 251-266. The microclimate of plastic shelters used for vegetable production.

197 Shakhov, A. A. & Golubkova, B. M. (1960). *Dokl. Akad. Nauk*

S.S.S.R., **135:** 486-8. [An electron microscope study of chloroplasts of plants with root systems kept at low temperatures.]

198 Shanks, J. B. & Laurie, A. (1949). *Proc. Am. Soc. hort. Sci.*, **53:** 473-488. A progress report of some rose root studies.

199 Shanks, J. B. & Laurie, A. (1949). *Proc. Am. Soc. hort. Sci.*, **54:** 495-499. Rose root studies: some effects of soil temperature.

200 Shanks, R. E. (1956). *Ecology*, **37:** 1-7. Altitudinal and microclimatic relationships of soil temperature under natural vegetation.

201 Shtrausberg, D. V. (1955). *Soviet Pl. Physiol.*, **5:** 226-232. The assimilation of nutritive elements by plants in the polar region under various temperature conditions.

202 Sinclair, J. G. (1922). *Mon. Weath. Rev. U.S. Dep. Agric.*, **50:** 142-144. Temperatures of the soil and air in a desert.

203 Smith, A. (1927). *Hilgardia*, **2:** 385-397. Effect of mulches on soil temperatures during the warmest week in July, 1925.

204 Smith, A. (1929). *Hilgardia*, **4:** 77-112. Daily and seasonal air and soil temperatures at Davis, California.

205 Smith, A. (1929). *Hilgardia*, **4:** 241-272. Comparisons of daytime and night-time soil and air temperatures.

206 Smith, A. (1931). *Hilgardia*, **6:** 159-201. Effect of paper mulches on soil temperature, soil moisture and yields of certain crops.

207 Smith, A. (1939). *Proc. Soil Sci. Soc. Am.*, **4:** 41-50. Value of mean and average soil and air temperatures.

208 Smith, E. M. (1959). *Ky Fm Home Sci.*, **5**(3): 4-6. Mechanising tobacco production.

209 Smith, G. D., Newhall, F., Robinson, L. H. & Swanson, D. (1964). *Tech. Pap. U.S. Dep. Agric. Soil Conserv. Serv.*, No. 1 CS, 1014, 14 pp. Soil temperature regimes_their characteristics and predictability.

210 Smith, G. E. P. (1936). *Agriculture, Lond.*, **17:** 383-385. Control of high soil temperature.

211 Smith, G. E. P., Kinnison, A. F. & Carns, A. G. (1931). *Tech. Bull. Ariz. agric. Exp. Stn*, **37:** 413-591. Irrigation investigations in young grapefruit orchards on the Yuma Mesa.

212 Smoliak, S. & Johnston, A. (1968). *Can. J. Pl. Sci.*, **48:** 119-127. Germination and early growth of grasses at four root-zone temperatures.

213 Snow, L. M. (1905). *Bot. Gaz.*, **40:** 12-48. The development of root hairs.

214 Sosebee, R. E. & Herbel, C. H. (1969). *Agron. J.*, **61:** 621-624. Effects of high temperatures on emergence and initial growth of range plants.

215 Stewart, G. R., Thomas, E. C. & Horner, J. (1926). *Soil Sci.*, **22:** 35-51. Some effects of mulching paper on Hawaiian soils.

216 Stuckey, I. H. (1942). *Pl. Physiol.*, **17:** 116-122. Influence of soil temperature on the development of Colonial bent grass.

217 Swezey, G. D. (1903). *A. R. Neb. agric. Exp. Stn*, No. 16, pp. 95-102. Soil temperatures at Lincoln, Nebraska, 1888-1902.

218 Tagawa, T. (1937). *J. Fac. Agric. Hokkaido imp. Univ.*, **39:** 271-296. The influence of the temperature of the culture water on the water absorption by the root and on the stomatal aperture.

219 Takatori, F. H., Lippert, L. F. & Whiting, F. L. (1964). *Proc. Amer. Soc. hort. Sci.*, **85:** 532-540. The effect of petroleum mulch and polyethylene films on soil temperature and plant growth.

220 Takeshima, H. (1964). *Proc. Crop Sci. Soc. Japan.*, **32:** 319-324. Studies on the effects of soil temperature on rice plant growth. 3. Effects of root temperature upon water and nutrient absorption at different stages and in alternating temperature.

221 Thompson, H. C. & Platenius, H. (1931). *Proc. Am. Soc. hort. Sci.*, **28:** 305-308. Results of mulch paper experiments with vegetable crops.

222 Thorskrud, J. (1965). *Yrkesfruktdyrking*, **1:** 1-16. Dyrkingsforsøk medjordbaer. VI. Forsøk med svart plastfolie til jorddekking.

223 Tothill, J. C. (1969). *Aust. J. Bot.*, **17:** 269-275. Soil temperatures and seed burial in relation to the performance of *Heteropogon contortus* and *Themeda australis* in burnt native woodland pastures in eastern Queensland.

224 Toumey, J. W. & Neethling, E. J. (1924). *Bull. Sch. For. Yale Univ.*, No. 11, 63 pp. Insolation as a factor in the natural regeneration of certain conifers.

225 Tsunoda, K. (1964). *Bull. nat. Inst. agric. Sci., Tokyo, Ser. A*, **11:** 75-174. Studies on the effects of water-temperature on the growth and yield of rice plants.

226 Unger, P. W. & Danielson, R. E. (1967). *Agron. J.*, **59:** 143-146. Water relations and growth of beans *(Phaseolus vulgaris* L.) as influenced by nutrient solution temperatures.

227 Vinokur, R. L. (1957). *Soviet Pl. Physiol.*, **4:** 268-273. Influence of temperature of the root environment on root activity, transpiration and photosynthesis of leaves of lemon.

228 Voth, V. & Bringhurst, R. S. (1961). *Proc. Amer. Soc. hort. Sci.*, **78:** 275-280. Pruning and polyethylene mulching of summer-planted strawberries in Southern California.

229 Voth, V. & Bringhurst, R. S. (1962). *Calif. Agric.*, **16**(2): 14-15. Early mulched strawberries.

230 Walker, J. C. (1926). *Phytopathology*, **16:** 697-710. The influence of soil temperature and soil moisture upon white rot of *Allium*.

231 Wallace, A. (1957). *Soil Sci.*, **83:** 407-411. Influence of soil temperature on cation uptake in barley and soybeans.

232 Wallace, A., Romney, E. M., Hale, V. Q. & Hoover, R. M. (1969). *Agron. J.*, **61:** 567-568. Effects of soil temperature and zinc application on yields and micronutrient content of four crop species grown together in a glasshouse.

233 Weaver, J. E. & Rowlands, N. W. (1952). *Bot. Gaz.*, **114:** 1-19. Effects of excessive natural mulch on development, yield, and structure of native grassland.

234 Weese, A. O. (1924). *Illinois biol. Monogr.,* **9**(4): 7-93. Animal ecology of an Illinois elm-maple forest.

235 Werner, H. C. (1933). *Bull. Neb. agric. Exp. Stn,* No. 278, 44 pp. Soil management experiments with vegetables.

236 Wharton, M. F. & Hobart, C. (1931). *Tech. Bull. Ariz. agric. Exp. Stn.,* **33:** 283-303. Studies in lettuce seedbed irrigation under high temperature conditions.

237 White, D. G., Pagan, C. & Manguel, J. C. (1947). *Trop. Agric., Trin.,* **24:** 131-136. The effects of mulching *Derris elliptica.*

238 White, P. R. (1937). *Pl. Physiol.,* **12:** 183-190. Seasonal fluctuations in growth rates of excised tomato root tips.

239 Whitfield, C. J. (1932). *Bot. Gaz.,* **94:** 183-196. Ecological aspects of transpiration. 2. Pikes Peak and Santa Barbara regions: edaphic and climatic aspects.

240 Wierenga, P. J. & Hagan, R. M. (1966). *Calif. Agric.,* **20:** 14-16. Effects of cold irrigation water on soil temperature and crop growth.

241 Wilcox, G. E., Martin, G. C. & Langston, R. (1962). *Proc. Am. Soc. hort. Sci.,* **80:** 522-529. Root zone temperature and phosphorus treatment effects on tomato seedling growth in soil and nutrient solutions.

242 Williams, C. B. (1923). *Bull. Minist. Agric. Egypt tech. sci. Serv.,* No. 29, 20 pp. A short bio-climatic study in the Egyptian desert.

243 Woodroof, J. G. (1934). *J. agric. Res.,* **49:** 511-530. Pecan root growth and development.

244 Wort, D. J. (1940). *Pl. Physiol.,* **15:** 335-342. Soil temperature and growth of Marquis wheat.

245 Zhurbitzky, Z. I. & Shtrausberg, D. V. (1958). In *Radioisotopes in scientific research.* Ed. R. C. Extermann, Vol. 4, pp. 270-285. London: Pergamon. The effect of temperature on the mineral nutrition of plants.

Certain aspects of these studies are reviewed in:

Cooper, A. J. (1972). Root temperature and plant growth. *A.D.A.S. quart. Rev.,* No. 6, pp. 12-19.

SPECIES INDEX

In this index each reference to a species appears once only – either under a scientific name or a common name where this is in general use or was so used in the original publication.